ADVANCES
IN CHILD DEVELOPMENT
AND BEHAVIOR

DEVELOPMENTAL DISORDERS AND
INTERVENTIONS

VOLUME 39

Contributors to This Volume

Paula J. Clarke

Kim M. Cornish

Darren L. Dunning

Susan E. Gathercole

David C. Geary

Kylie M. Gray

Lisa M. Henderson

Joni Holmes

Glynis Laws

Melanie A. Porter

Deborah M. Riby

Nicole J. Rinehart

Emma Truelove

ADVANCES
IN
CHILD DEVELOPMENT
AND
BEHAVIOR

DEVELOPMENTAL DISORDERS AND

INTERVENTIONS

edited by

Joni Holmes
Department of Psychology
Northumbria University
Newcastle Upon Tyne
United Kingdom

VOLUME 39

AMSTERDAM • BOSTON • HEIDELBERG • LONDON
NEW YORK • OXFORD • PARIS • SAN DIEGO
SAN FRANCISCO • SINGAPORE • SYDNEY • TOKYO
Academic Press is an imprint of Elsevier

Academic Press is an imprint of Elsevier
32 Jamestown Road, London NW1 7BY, UK
Radarweg 29, PO Box 211, 1000 AE Amsterdam, The Netherlands
30 Corporate Drive, Suite 400, Burlington, MA 01803, USA
525 B Street, Suite 1900, San Diego, CA 92101-4495, USA

First edition 2010

Recognizing the importance of preserving what has been written, Elsevier prints
its books on acid-free paper whenever possible.

Library of Congress Cataloging-in-Publication Data
A catalogue record for this book is available from the Library of Congress

British Library Cataloguing in Publication Data
A catalog record for this book is available from the British Library

ISBN: 978-0-12-374748-8
ISSN: 0065-2407 (Series)

For information on all Academic Press publications
visit our website at elsevierdirect.com

Printed in the United States of America

10 11 12 13   9 8 7 6 5 4 3 2 1

Contents

Poor Working Memory: Impact and Interventions

JONI HOLMES, SUSAN E. GATHERCOLE, AND DARREN L. DUNNING

Mathematical Learning Disabilities

DAVID C. GEARY

The Poor Comprehender Profile: Understanding and Supporting Individuals Who Have Difficulties Extracting Meaning from Text

PAULA J. CLARKE, LISA M. HENDERSON, AND EMMA TRUELOVE

Contributors

PAULA J. CLARKE
School of Education, University of Leeds, Leeds, United Kingdom
KIM M. CORNISH
Centre for Developmental Psychiatry and Psychology, School of Psychology and Psychiatry, Faculty of Medicine, Nursing and Health Sciences, Monash University, Melbourne, Australia
DARREN L. DUNNING
Department of Psychology, University of York, York, United Kingdom
SUSAN E. GATHERCOLE
Department of Psychology, University of York, York, United Kingdom
DAVID C. GEARY
Department of Psychological Sciences, University of Missouri, Columbia, Missouri, USA
KYLIE M. GRAY
Centre for Developmental Psychiatry and Psychology, School of Psychology and Psychiatry, Faculty of Medicine, Nursing and Health Sciences, Monash University, Melbourne, Australia
LISA M. HENDERSON
Department of Psychology, University of York, Heslington, York, United Kingdom
JONI HOLMES
Department of Psychology, Northumbria University, Newcastle upon Tyne, United Kingdom
GLYNIS LAWS
Department of Experimental Psychology, University of Bristol, Bristol, United Kingdom
MELANIE A. PORTER
Department of Psychology, Macquarie University, Australia
DEBORAH M. RIBY
School of Psychology, Newcastle University, United Kingdom
NICOLE J. RINEHART
Centre for Developmental Psychiatry and Psychology, School of Psychology and Psychiatry, Faculty of Medicine, Nursing and Health Sciences, Monash University, Melbourne, Australia
EMMA TRUELOVE
Department of Psychology, University of York, Heslington, York, United Kingdom

Preface

This thematic collection provides an overview of contemporary research into neurodevelopmental disorders of learning. In each chapter the authors provide a synopsis of the cognitive, social, and emotional profiles of children with different developmental disorders and, where appropriate, tease out the neural and genetic underpinnings of these problems. Due to the far-reaching negative impact of these difficulties, the authors provide an in-depth critique of different approaches to intervention. In most cases they consider a predominantly cognitive approach (chapters on poor working memory, mathematical disabilities, poor comprehenders and Downs syndrome), although in the later chapters the authors also consider pharmacological, social, and behavioral methods (chapters on Williams syndrome and Fragile X syndrome). The emphasis on intervention in each chapter is both timely and highly relevant to academics and professionals working in education, health, and medicine who are seeking to improve the quality of life, social functioning, and learning outcomes of the vast number of children affected by these disorders.

In Chapter 1, we are introduced to children with poor working memory. These children are not commonly diagnosed, yet they represent a substantial minority of children who are struggling at school who do not have recognized special educational needs or receive tailored support. In this chapter the authors discuss three main approaches to alleviating the difficulties faced by children with poor working memory: a classroom intervention, strategy training, and direct working memory training. The latter of these approaches is shown to be extremely effective in boosting working memory, not only in children with poor working memory, but also in children with other disorders that are associated with low memory function, such as ADHD and dyslexia. These findings hold the promise for improving the long-term educational outcomes of children with a wide range of disorders that are associated with poor working memory function.

In Chapters 2 and 3, we focus on two cognitive disorders of learning; reading and mathematical difficulties. Both disorders are multifaceted and encompass different profiles. In the case of mathematical difficulties, Geary distinguishes between low achieving children and those with mathematical learning disabilities: children who score at or below the 10th percentile on standardized mathematics achievement tests for at least two consecutive academic years are categorized as having mathematical learning disabilities; children scoring between the 11th and the 25th percentiles across two consecutive years are categorized as low achievers. Geary considers both profiles throughout his chapter. In contrast, Clarke and colleagues focus on one subgroup of children with reading difficulties: poor comprehenders. These are children who despite being able to read accurately experience difficulties extracting meaning from text. This profile is not yet widely recognized by practitioners and has received considerably less attention than other reading difficulty profiles such as dyslexia. However, because it affects a significant minority of children who experience difficulty making the transition from learning to read to reading to learn, this chapter aims to raise awareness of its profile.

Due to the complex nature of these difficulties, both Geary and Clarke *et al.* pay particular attention to the definitions and criteria used to accurately assess and identify children with poor mathematical or comprehension skills. Following the tradition in both fields to identify the core deficits associated with the disorders and the cognitive mechanisms which may mediate the expression of these deficits, both chapters also focus on identifying cognitive explanations. As we learn, each is characterized by specific deficits, either in mathematical cognition (counting, arithmetic, and number knowledge) or in language (semantics, grammar, pragmatics, and inferencing), as well as deficits in domain-general mediators that include working memory and processing speed in children with mathematical disabilities and working memory, inhibition, and metacognitive skills in children who struggle to extract meaning from text.

In line with these profiles, both Geary and Clarke *et al.* overview interventions that target precise areas of cognitive weaknesses in their respective groups, alongside those that have a broader focus on cognitive impairment. Geary, for example, suggests that interventions targeting either specific mathematical cognition deficits (e.g., Fuchs *et al.*, 2010) or fundamental mechanisms underlying these deficits, such as working memory (e.g., Holmes, Gathercole, & Dunning, 2009), may be effective for children with mathematical difficulties. Likewise, in their chapter, Clarke and colleagues advocate the implementation of targeted interventions for language, such as inference and vocabulary training, as well as multicomponent packages that tackle both general mediators such as metacognition and the aspects of language that are specifically impaired in poor comprehenders (e.g., Clarke, Snowling, Truelove, & Hulme, 2010).

The latter half of this collection focuses on three genetic disorders of learning: Down syndrome, Williams Syndrome, and Fragile X syndrome.

Each disorder is characterized by a distinctive genetic profile: children with Down syndrome have a trisomy of chromosome 21; Williams syndrome is characterized by the deletion of 25 genes on chromosome 7; and Fragile X is caused by a single gene being switched off on the X chromosome. Although the etiology of these neurodevelopmental disorders is better understood than the cognitive difficulties discussed earlier in this volume, the challenge for researchers to understand the phenotypes of each condition and develop effective research-based interventions remains the same.

In Chapter 4, Laws provides a detailed analysis of the cognitive profile of children with Down syndrome. Akin to children with Fragile X syndrome, who are discussed in Chapter 6, these children are characterized by moderate to severe impairments in intellectual functioning. Both groups also have weaknesses in language and speech production, although the expression of these deficits differs across disorders. While marked impairments in speech production are common in children with Down syndrome, children with Fragile X syndrome have difficulties in fluency that are characterized by repetitive and impulsive speech. Fragile X syndrome is also accompanied by difficulties with grammar and pragmatics, whereas Down syndrome is characterized by compromised expressive language skills and deficits in the use and understanding of morphology and syntax. Unlike these groups, children with Williams syndrome, who are the focus of Chapter 5, have a relative proficiency in overall language ability and less severe intellectual impairments (typically in the mild to moderate range).

Laws explains that children with Down syndrome have substantial impairments in verbal short-term memory, but relatively intact visuo-spatial memory skills. This contrasts with the profiles of children with Fragile X syndrome and Williams syndrome, whose verbal memory skills are relatively strong in comparison with their visuo-spatial skills. Children with these two disorders also evidence impairments in attention and executive control that increase with age.

Children with Williams syndrome are described as hypersociable with an atypically increased desire to approach unfamiliar people and gaze at people and faces. In contrast, children with Fragile X are described as having a desire for social contact that is typically coupled with socially anxious and gaze-avoidant behavior. Both disorders, however, are associated with other maladaptive behaviors. Riby, Cornish, and colleagues indicate that approximately 64% of children with Williams syndrome and 67–84% of children with Fragile X syndrome meet the criteria for ADHD respectively. Depressive and anxiety disorders are also common among individuals with Williams syndrome and 24–33% of Fragile X children meet the criteria for a clinical diagnosis of autism, with almost all displaying some autistic features.

Treatment options discussed for the genetic syndromes focus on the behavioral symptoms and areas of cognitive weakness associated with the disorders. In Chapter 4, Laws evaluates the impact of reading as an intervention for the oral language, verbal short-term memory, and speech production impairments prevalent in children with Down syndrome. In a comparison of the receptive vocabulary, verbal short-term memory, and speech production of nonreaders and emerging readers with Down syndrome, she found no evidence that these areas of relative weakness improved over time in better readers. Laws cites several reasons for these negative outcomes and suggests that teaching phonological awareness to children with Down syndrome, either independently or as part of an integrated approach to literacy teaching, could have a positive impact on oral language and verbal memory. Riby and Porter suggest that a phonic-based approach to reading might also benefit children with Williams syndrome, although they exercise caution in advocating a universal approach to intervention with children with Williams syndrome due to their heterogeneous language profiles.

A broader perspective on approaches to intervention is provided in the final two chapters of this volume. Riby and Porter outline potential remediation strategies for children with Williams syndrome that emphasize the use of areas of cognitive strength to intervene areas of weakness, such as using verbal mediation or music therapy to promote the learning of nonverbal and mathematical concepts. They also suggest that other therapies, such as speech and language therapy or cognitive behavior therapy, might be useful for treating the language and social difficulties associated with the disorder respectively. The latter suggestion mirrors Cornish *et al.*'s review of intervention approaches for Fragile X syndrome, which principally advocates the use of treatment options that target problem behaviors that also occur in other developmental disorders and for which treatment is well established. Although both Riby and Cornish recognize the need to tailor interventions both to specific disorders and to the individual profiles of children within each disorder, they agree that pharmacological and behavioral approaches used to treat children with ADHD and autism might be effective, with modifications, in alleviating problem behaviors and improving cognitive functioning in children with Williams syndrome and Fragile X syndrome.

In summary, the majority of children with developmental disorders discussed here are characterized by both disorder-specific (e.g., face processing or comprehension impairments) and disorder-general impairments (e.g., slow processing speed or poor working memory) that are relative to their overall level of functioning. The shared areas of weakness between syndromes raise the possibility that existing interventions used widely for one disorder could, with modifications, alleviate the symptoms of other disorders. However, the variability both within and across these pathologies, as well as the different developmental pathways that lead to the expression of these common characteristics, may prevent this kind of one-size-fits-all approach.

Clearly then, the challenges for the future are to continue with the fine-grained assessment of the multiple levels of explanation of developmental disorders, while incorporating cross-syndrome perspectives to allow us to develop both a deeper understanding of disorder-specific profiles and theoretically driven, evidence-based approaches to intervention that may, or may not, be syndrome-specific.

Joni Holmes
Department of Psychology,
Northumbria University,
Newcastle Upon Tyne, United Kingdom

REFERENCES

Clarke, P. J., Snowling, M. J., Truelove, E., & Hulme, C. (2010). Ameliorating children's reading comprehension difficulties: A randomised controlled trial. *Psychological Science*, (online early view).

Fuchs, L. S., Powell, S. R., Seethaler, P. M., Cirino, P. T., Fletcher, J. M., Fuchs, D., et al. (2010). The effects of strategic counting instruction, with and without deliberate practice, on number combination skill among students with mathematics difficulties. *Learning and Individual Differences, 20*, 89–100.

Holmes, J., Gathercole, S. E., & Dunning, D. L. (2009). Adaptive training leads to sustained enhancement of poor working memory in children. *Developmental Science, 12*, F9–F15.

POOR WORKING MEMORY: IMPACT AND INTERVENTIONS

Joni Holmes,* Susan E. Gathercole,† and Darren L. Dunning†

* DEPARTMENT OF PSYCHOLOGY, NORTHUMBRIA UNIVERSITY, NEWCASTLE UPON TYNE, UNITED KINGDOM
† DEPARTMENT OF PSYCHOLOGY, UNIVERSITY OF YORK, YORK, UNITED KINGDOM

I. Introduction

Poor working memory is not yet a recognized developmental disorder. However, children with poor working memory function are at very high risk of educational underachievement (e.g., Gathercole & Alloway, 2008) and it has recently been suggested that deficits in working memory might be a cognitive phenotype for children who make slow progress at school, but who do not have general learning difficulties (Gathercole, 2010). Moreover, working memory impairments are associated with a wide range of developmental disorders of learning, including attention-deficit hyperactivity disorder (ADHD), dyslexia, specific language impairment (SLI),

1

Down syndrome, and reading and mathematical difficulties (see Alloway & Gathercole, 2006). The early recognition of working memory difficulties and the provision of effective educational support and targeted intervention are therefore paramount to improving the long-term outcomes for a vast number of children.

So, what is working memory and how can we identify children with *poor* working memory? The term working memory describes our ability to hold in mind and manipulate information for brief periods of time during complex cognitive activities. There are several theoretical models of working memory which differ in their views of the nature, structure, and function of the system (for a review, see Conway, Jarrold, Kane, & Towse, 2007; Miyake & Shah, 1999). The primary distinction between these models is whether working memory is conceived as a distinct entity (e.g., Baddeley, 2000; Baddeley & Hitch, 1974) or a limited capacity process of controlled attention that acts to activate existing representations in long-term memory which then become the current contents of working memory (e.g., Anderson & Lebiere, 1998; Barrouillet, Bernardin, & Camos, 2004; Cowan, 2005; Engle, Kane, & Tuholski, 1999; Kintsch, Healy, Hegarty, Pennington, & Salthouse, 1999). Most accounts of working memory distinguish between the storage-only capacities of short-term memory (STM) and the storage and processing functions of working memory. They also agree that the capacity of working memory is limited, meaning there is an upper limit to the amount of information that can be stored and processed at any given moment in time.

According to one of the most widely accepted models (Baddeley, 2000; Baddeley & Hitch, 1974), working memory is a multicomponent system. There are two domain-specific STM stores, the phonological loop and the visuospatial sketchpad, that are specialized for the temporary maintenance of verbal and visual and spatial information, respectively. These are governed by a domain-general central executive system linked to attentional control, responsible for holding and manipulating information from long-term memory, coordinating performance on dual tasks, switching between retrieval strategies, and inhibiting irrelevant information (e.g., Baddeley, 1996; Baddeley, Emslie, Kolodny, & Duncan, 1998; Engle & Kane, 2004; Engle *et al.*, 1999; Kane, Conway, Hambrick, & Engle, 2007, Kane & Engle, 2000; Kane, Hambrick, & Conway, 2005). The fourth component, the episodic buffer, is responsible for integrating information from the subcomponents of working memory and long-term memory (Baddeley, 2000).

An individual's working memory capacity can be assessed using both simple and complex span tasks. Simple span tasks, also known as STM tasks, involve the storage of either verbal or visuospatial material. These assess the phonological loop and visuospatial sketchpad components of the Baddeley and Hitch (1974) working memory model, respectively. Complex memory span, or "working memory," tasks impose significant demands on both storage and processing, and require the support of both the central executive, plus the respective storage components of working memory. Thus, verbal working memory tasks such as reading and listening span (Daneman & Carpenter, 1980; Daneman & Merickle, 1996) rely on both the central executive and the phonological loop, whereas working memory tasks involving visuospatial material draw upon the central executive and the visuospatial sketchpad (Alloway, Gathercole, & Pickering, 2006; Kane *et al.*, 2004).

There are currently two standardized test batteries designed to assess children's working memory capacities, the Automated Working Memory Assessment (Alloway, 2007) and the Working Memory Test Battery for Children (Pickering & Gathercole, 2001). Both provide multiple assessments of the different aspects of short-term and working memory and, because age norms are available, can be used to identify children who have poor working memory for their age. In our own research, we typically classify children as having poor working memory skills if their standard scores on two separate tests of working memory are at least one standard deviation below the mean (bottom 15th centile, standard scores < 86). It is these children who we know are at substantial risk of poor educational progress. When diagnosing a child with poor working memory, it is important to take into account potential non-memory causes of low test performance. For example, factors such as speech-motor dysfunction and hearing or sight problems might give rise to inaccurate encoding or recall, which will affect performance negatively. The Working Memory Rating Scale (Alloway, Gathercole, & Kirkwood, 2008) can also be used to identify children with working memory impairments. This is a behavioral rating scale consisting of 22 statements, which are rated by a child's teacher. It provides a quick and efficient method for the early identification of working memory problems in a school setting.

II. Cognitive Profile

Poor working memory is associated with a wide range of cognitive difficulties that primarily relate to learning, but which we have recently seen extend to other executive functions including planning, problem solving, and sustained attention (Gathercole *et al.*, 2010). Low working memory is also related to below-average intelligence (IQ) and deficits in working memory are key markers of a number of developmental disorders of learning. Each of these issues is considered in the following sections.

A. POOR WORKING MEMORY AND LEARNING IMPAIRMENTS

Working memory skills are highly associated with children's abilities to learn in key academic domains such as reading, mathematics, and science (Gathercole & Pickering, 2000; Geary, Hoard, Byrd-Craven, & De Soto, 2004; Holmes & Adams, 2006; Jarvis & Gathercole, 2003; Swanson & Saez, 2003). Children's performances on tests of working memory are also significantly associated with the developmental precursors of reading, writing, and mathematics as children commence formal schooling (Gathercole, Brown, & Pickering, 2003) and are excellent prospective indicators of academic performance, predicting children's attainment on National achievement tests at 7, 11, and 14 years of age (Gathercole, Pickering, Knight, & Stegmann, 2004; Jarvis & Gathercole, 2003).

Working memory is used to process and store information during complex and demanding activities (Just & Carpenter, 1992). It therefore supports many activities children routinely engage in at school. Imagine, for example, attempting to read and comprehend a passage of text. The process of reading sentences, holding them in mind, and integrating the information to uncover the meaning relies heavily on the ability to simultaneously process and store information over the short term. Similarly, solving a mathematics problem, such as multiplying two numbers together, involves the temporary storage of digits whilst simultaneously retrieving learned rules of multiplication and stored number facts from long-term memory.

Beyond support for everyday mental activities, working memory also provides vital support for learning across the school years. Children with poor working memory characteristically underachieve at school (Gathercole & Alloway, 2008) and, conversely, children who underperform at school typically have working memory impairments. Gathercole and colleagues found that 41% of children who achieved below-average scores on National English tests at 6 and 7 years of age had working memory scores in the deficit range, as did 52% of children who achieved the same low levels in National Mathematics tests at this age. Similar proportions were reported for adolescents who scored in the below-average range on National tests in English, Mathematics, and Science at 14 years of age, with more severe working memory impairments associated with below-average performance in Science and Mathematics. Across both age groups, STM scores for low achievers did not differ significantly to average and high achievers, suggesting it is working memory rather than STM that limits learning opportunities (Gathercole *et al.*, 2004). Overall, these data show that the incidence of poor working memory is more than three times higher in low achievers compared to the normal school population, in which approximately only 16% would be expected to show working memory deficits.

At the extreme end of the school distribution, distinctive working memory profiles characterize children with special educational needs (SEN). Pickering and Gathercole (2004) reported that children with general learning difficulties in English/literacy and mathematics were six times more likely to have both poor verbal and visuospatial STM and verbal working memory scores than children without SEN. Children with SEN specifically related to language were also impaired on tasks that measured short-term and working memory, although their STM deficits were limited to the verbal domain. Conversely, children with noncognitive SEN such as behavioral problems had working memory skills in the normal range. These distinctive profiles suggest that children with SEN have deficits in working memory that compromise their educational progress.

The severity of a child's SEN in literacy or mathematics is associated with the severity of their memory impairment. Children whose needs are of the greatest severity, indexed by requiring additional resources to support their learning, show commensurately greater cognitive deficits than children with milder learning difficulties. Their deficits are particularly marked in tasks measuring verbal working memory (Alloway, Gathercole, Adams, & Willis, 2005).

Studies of children identified solely on the basis of poor working memory, rather than poor educational progress or SEN, show that they are very likely to struggle at school. Gathercole and Alloway (2008) recently examined the academic profiles of a large group of children with poor working memory. Over 300 children aged 5 or 6 and 9 or 10 with scores in the bottom 10th centile were identified. Of these children, 75% of the 5 and 6 year olds and 83% of the 9 and 10 year olds had difficulties in both reading and mathematics. An additional 5% of the 5 and 6 year olds were struggling in mathematics only. These figures clearly illustrate impaired rates of learning in children with poor working memory.

These children also struggle to successfully complete a range of tasks that are designed to aid learning at school. Common classroom activities that require large amounts of information to be held in mind are particularly challenging for children with poor working memory. One of the most crucial aspects of classroom learning is following spoken instructions given by the teacher, and this is particularly difficult for children with small working memory capacities. Teacher instructions are often multistep, directing children where they or their classroom objects should be, contain vital information about learning activities, or relate to a sequence of actions that must be carried out. To perform these actions, children must be able to remember the different parts of the instruction whilst carrying out the various steps to complete the action successfully. Children with poor working memory typically either carry out the first command of a multistep instruction, skip straight to the last step, or simply abandon the task all together as they are unable to remember all the necessary parts of the sequence (Gathercole & Alloway, 2008; Gathercole, Lamont, & Alloway, 2006). Children with low working memory have also been shown to be poorer in laboratory tasks that involve carrying out the actions described by a multistep instruction and remembering the content of the instruction (Gathercole, Durling, Evans, Jeffcock, & Stone, 2008).

These children also experience difficulties in classroom activities that require information to be both processed and stored. For example, writing activities that require children to count the number of words in a sentence, write down the sentence, and then check that what they have written contains the same number of words as were originally counted require children to store the number of words in mind, remember the sentence, and also engage in the mentally challenging processing of writing down the sentence (Gathercole & Alloway, 2008).

Children with poor working memory also make characteristic errors in their classroom work. These include failing to keep track of their place in demanding and complex activities and mistakes in writing and counting. In their written work, they miss out letters, skip whole words, or blend parts of different words from different sentences. So, for example, if they were asked to copy the title of a piece of work "My Holiday" and the date "17th June," they might write down "My Holune." In counting tasks, they lose track of which numbers they are working with and where they are in the calculation. For example, if they were adding the numbers 28, 7, and 11, they might miss out the 7 and only add together the numbers 28 and 11. In most cases, children fail to self-correct which ultimately leads to failure (Gathercole & Alloway, 2008).

These kinds of activities, which are common place in the classroom, typically overload the working memory capacities of many children. This memory overload causes problems in following teacher instructions, remembering information vital to individual tasks, and keeping track in structured learning activities (Gathercole & Alloway, 2008; Gathercole *et al.*, 2006). Over time, these frequent missed learning opportunities amount to slow educational progress and poor academic attainment (Gathercole & Alloway).

B. LOW WORKING MEMORY AND IQ

There is considerable overlap between performance on tests of working memory and IQ, with correlations as high as 0.8 reported between working memory and reasoning tasks (Kyllonen & Christal, 1990) and similarly high levels of association observed between executive function and IQ tests (Miyake, Friedman, Shah, Rettinger, & Hegarty, 2001). One explanation is that the two domains are overlapping but dissociable and that the shared variance between working memory and IQ occurs because methods of assessing IQ are strongly influenced by working memory (e.g., Jurden, 1995). Carpenter, Just, and Shell (1990) first suggested that working memory capacity may be a main factor underpinning individual differences on the Raven's Progressive Matrices test, a commonly used nonverbal IQ test, in the early 1990s. Since then, working memory has been reported to predict performance on various tests of general IQ (Engle *et al.*, 1999; Kane *et al.*, 2004) and assessments of working memory are now an integral part of one of the most widely used IQ assessments, the Wechsler Intelligence Scales for Children IV (Wechsler, 2004).

There are, in fact, clear distinctions between IQ and working memory. First, working memory tasks assess different aspects of a single, well-understood memory system. IQ assessments, however, aggregate performance across a wide range of abilities, which includes language and nonverbal reasoning. The two domains are also separate in the extent to which a child's prior knowledge and experiences contribute to performance. Verbal IQ assessments rely heavily on a child's knowledge of words and language, which are influenced by background factors such as their home language environment and level of education (Brooks-Gunn, Klebanov, & Duncan, 1996; Hoff & Tian, 2005; Huttenlocher, Haight, Bryk, Seltzer, & Lyons, 1991). Working memory tasks, however, are equally unfamiliar to all children, conferring no obvious advantages or disadvantages to children from different backgrounds. Studies have shown that children from minority ethnic groups and poor economic circumstances score poorly on tests of vocabulary knowledge but not on measures of either verbal STM (Campbell, Dollaghan, Needleman, & Janosky, 1997; Ellis Weismer *et al.*, 2000; Engel, Santos, & Gathercole, 2008; Santos & Bueno, 2003) or verbal working memory (Engel *et al.*, 2008).

Working memory and IQ also make distinct contributions to learning. Working memory predicts unique variance in children's attainment above and beyond what can be predicted by IQ (Gathercole, Alloway, Willis, & Adams, 2006). Furthermore, associations between working memory and attainment persist after differences in performance IQ or verbal IQ have been statistically controlled both in children with and without learning difficulties (e.g., Cain, Oakhill, & Bryant, 2004; Siegel & Ryan, 1989; Swanson & Sachse-Lee, 2001). For these reasons, working memory tests appear to provide culture-fair indices of a child's cognitive potential.

Despite differences between working memory and IQ, children with poor working memory typically have low range IQ, and vice versa. In a recent comparison of the cognitive profiles of 50 children with low working memory (standard scores below 86 on two tests of working memory) and 50 age-matched children with average working memory (standard scores over 90 on two working memory assessments), we found significant group differences in both verbal and performance IQ. In both cases, the average working memory group had significantly higher scores than the low memory group (see Figure 1).

Of the low working memory group, 14% had verbal IQ scores in the extremely low range (standard scores below 70), 30% scored in the poor range (standard scores in the 70–80 range), 54% in the average range (scores between 86 and 115), and a further 2% scored above average (standard scores in excess of 116). For performance IQ, 6% scored in the extremely low range, 50% in the poor range, and 44% in the average range. Overall, these data show that the majority of children with poor working memory have IQ scores in the poor average range, with very few in the extremely low IQ category (Gathercole *et al.*, 2010).

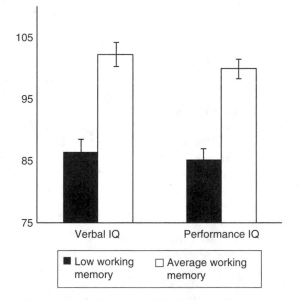

Fig. 1. Verbal and performance IQ scores of 50 children with poor working memory and 50 age-matched children with average working memory, from Gathercole et al. (2010).

C. WORKING MEMORY DEFICITS AND PERVASIVE EXECUTIVE FUNCTION IMPAIRMENTS

Working memory is one of several executive functions that support flexible goal-directed behavior. Others include inhibition, switching, planning, and problem solving (Pennington & Ozonoff, 1996). Inhibition involves controlling an overlearned or prepotent response (Stroop, 1935); switching/shifting is involved in changing between mental sets, multiple tasks, or from one situation, or aspect of a situation, to another (e.g., Monsell, 1996); planning involves setting goals, developing appropriate steps ahead of time, and anticipating future events; and problem solving allows us to initiate behavior to solve tasks by engaging in flexible thinking.

The extent to which problems in working memory extend to other executive functions is not at present well understood. In children as in adults, individual differences studies have distinguished working memory from inhibitory control and set-shifting behavior (Miyake *et al.*, 2000; St. Clair-Thompson & Gathercole, 2006). However, behavioral ratings of children with poor working memory include executive function difficulties such as monitoring the quality of their own work and generating new solutions to problems (Gathercole, Alloway, *et al.*, 2008).

To investigate whether the working memory problems faced by children with poor working memory are part of a more pervasive pattern of impaired executive function, we recently assessed 50 children with working memory deficits, and a comparison group of 50 age-matched children with average working memory, on a range of tests of executive function (Gathercole *et al.*, 2010). All children completed standardized tests of cognitive inhibitory control, set-shifting, planning, motor inhibition, and sustained attention. Where appropriate, measures of basic information processing were taken for comparison with higher level executive control. Performance for both groups on the primary higher level tasks is summarized in Table I.

Overall, we found that children with poor working memory had difficulties that extended to other executive functions, but which did not reflect higher level executive impairments in set-shifting or inhibition. Children with poor working memory were characterized by significantly higher error rates on a set-shifting task and an inhibition task, a greater number of rule violations on a planning task, and a higher incidence of omissions on a sustained attention task in comparison to the average working memory group. They also had significantly slower completion times on the inhibition task and were significantly impaired relative to the comparison group on measures of motor inhibition and problem

Table I

Mean Executive Function Scores for Low and Average Working Memory Children,
from Gathercole et al. (2010)

Measure		Low WM		Average WM	
		M	SD	M	SD
Set-shifting	Time	9.50	3.11	10.56	2.92
	Errors	30.17	21.62	45.51	18.31
Inhibition	Time	9.38	3.02	11.68	2.92
	Errors	44.30	27.37	65.00	25.07
Planning	Total	12.54	4.88	12.94	3.83
	Errors	38.60	39.33	61.54	41.11
Sustained attention	Omissions	40.56	29.40	23.36	21.21
	Commissions	74.40	74.36	48.04	51.65
Motor inhibition	Total	4.14	3.57	9.28	3.69

Note: Set-shifting, inhibition time scores, and planning and motor inhibition total scores are scaled scores ($M = 10$, $SD = 3$); set-shifting, inhibition, and planning error scores are cumulative centiles-lower scores indicate a higher number of errors; sustained attention scores are counts.

solving. However, their performance on the inhibition and set-shifting tasks was not disproportionately greater in conditions that required executive intervention compared to their performance on tasks that tapped the basic cognitive processes necessary for completion of the higher level task. For example, although they performed poorly on the Stroop-like inhibition task, their performance on this higher level task was not disproportionately poorer than on the basic color naming and word reading aspects of the task. Thus, their weaknesses in basic processing underpinned their impairments in the higher level executive tasks. These data show that children with low working memory have deficits that are manifest in a variety of cognitively demanding activities, but that they do not have selective impairments in high-level executive functions of inhibition or set-shifting.

In summary, there is evidence from these standardized cognitive assessments that executive function impairments that extend beyond working memory are present in these children. This is also strongly reflected in teacher ratings of classroom behavior, which point to problems in monitoring and planning, problem solving, and shifting (Gathercole, Alloway, *et al.*, 2008). Although they struggle in cognitive tasks designed to measure planning, problem solving, and sustained attention, children with poor working memory do not have specific deficits in inhibition and set-shifting (Gathercole *et al.*, 2010).

Working memory and executive difficulties might co-occur because limited working memory capacities constrain performance on tasks that explicitly require the storage and processing of information. For example, problems in planning and monitoring may result from the loss of task information and goals from working memory. It is also possible that the inattentive profile of children with poor working memory results from the loss of goal-critical information from working memory (this is discussed in detail later in this chapter). By this account, poor working memory function underpins deficits in a range of executive tasks. However, the issue of causality is one that needs further exploration as it is equally possible that executive function failures may be the cause rather than the consequence of poor working memory.

D. WORKING MEMORY AND DEVELOPMENTAL DISORDERS OF LEARNING

The cognitive profiles of children with poor working memory overlap with a number of developmental disorders of learning. These include reading and mathematical difficulties, dyslexia, SLI, Down syndrome, William's Syndrome, and ADHD.

Very low levels of performance on working memory tasks are common in children with specific difficulties in reading (Gathercole, Alloway, *et al.*, 2006; Pickering & Gathercole, 2004; Swanson, 1993, 2003). Verbal STM is significantly associated with reading development during the early years (Gathercole & Baddeley, 1993) and deficits in this component of the memory system are common among children with reading difficulties (Siegel & Ryan, 1989; Swanson & Siegel, 2001). Verbal working memory skills have also been found to be consistently associated with children's reading skills (e.g., de Jonge & de Jong, 1996; Engle, Carullo, & Collins, 1991) and explain unique variance in reading comprehension over and above verbal STM, word reading, and vocabulary knowledge (e.g., Cain *et al.*, 2004; Swanson & Jerman, 2007). Furthermore, impairments in complex span tasks that tap working memory extend across both the verbal and nonverbal domain, indicative of a modality-general impairment in working memory in poor readers (Chiappe, Hasher, & Siegel, 2000; de Jong, 1998; Gathercole, Alloway, *et al.*, 2006; Palmer, 2000; Swanson, 1993).

Individuals whose reading problems satisfy the more stringent criteria for dyslexia also perform below average on both short-term and working memory tasks in the verbal domain (Jeffries & Everatt, 2003, 2004). Children with SLI show the same pattern of highly specific deficits in the verbal domain, with severe impairments in both verbal STM (Archibald & Gathercole, 2006; Edwards & Lahey, 1998; Ellis Weismer,

Evans, & Hesketh, 1999; Gathercole & Baddeley, 1990; Montgomery, 1995) and verbal working memory (Archibald & Gathercole, 2007; Ellis Weismer *et al.*, 1999; Montgomery, 2000a, 2000b). It has been suggested that poor verbal storage skills underlie impairments in verbal working memory in this group (Archibald & Gathercole).

Children with mathematical difficulties also show signs of working memory deficits (Bull & Scerif, 2001; Geary, 1993; Mayringer & Wimmer, 2000; Passolunghi & Siegel, 2004; Siegel & Ryan, 1989; Swanson & Beebe-Frankenberger, 2004). These children typically perform poorly on measures of visuospatial STM and working memory (Gathercole & Pickering, 2000; Geary, Hoard, & Hamson, 1999; McLean & Hitch, 1999; Siegel & Ryan), but not on measures of verbal STM (McLean & Hitch; Passolunghi & Siegel).Working memory appears to play an important role in the development of counting, with children with poor working memory using primitive finger-counting strategies that have relatively low working memory demands (Geary *et al.*, 2004). Their continued use of these early strategies prevents them establishing networks of arithmetic facts in long-term memory, which are necessary to support the use of efficient retrieval-based strategies analogous to those used in adulthood (e.g., Hamann & Ashcraft, 1985; Kaye, 1986). Thus, poor working memory impedes the learning of number facts (Geary, 2004), the learning and efficiency of number transcoding (Camos, 2008; McLean & Hitch) and computational skills (Wilson & Swanson, 2001). It also causes difficulties in solving mathematical problems expressed in everyday language (Swanson & Sachse-Lee, 2001).

Impairments in working memory are also associated with a variety of genetic pathologies, including Down syndrome and William's syndrome. There is considerable evidence for marked deficits in verbal STM among children with Down syndrome (e.g., Jarrold, Baddeley, & Hewes, 1999). These children typically perform at age-appropriate levels on visuospatial STM tasks and do not appear to have deficits in working memory when compared to controls (Numminen, Service, Ahonen, & Ruoppila, 2001; Pennington, Moon, Edgin, Stedron, & Nadel, 2003). In marked contrast, children with William's syndrome have much stronger verbal STM than visuospatial STM skills (Jarrold, Baddeley, Hewes, & Phillips, 2001). This pattern of impairment is most likely related to the double dissociation between verbal and visual processing skills in William's syndrome.

Children with behavioral difficulties such as ADHD are also characterized by poor working memory function (Martinussen, Hayden, Hogg-Johnson, & Tannock, 2005; Willcutt, Doyle, Nigg, Faraone, & Pennington, 2005). Children with ADHD perform poorly on tests of visuospatial STM (Barnett *et al.*, 2001; Martinussen *et al.*; Mehta, Goodyear,

& Sahakian, 2004) and both verbal and visuospatial working memory tasks (Martinussen & Tannock, 2006; Martinussen *et al.*; McInnes, Humphries, Hogg-Johnson, & Tannock, 2003; Roodenrys, 2006; Willcutt, Doyle, *et al.*, 2005). Their verbal STM appears to be relatively preserved, suggesting that verbal storage problems are not fundamental features of the disorder (e.g., Martinussen & Tannock). Our own data from a sample of 83 children aged 8–11 years with a clinical diagnosis of combined type ADHD concur with this pattern of impairment. We found that whilst verbal STM was relatively intact in this sample, visuospatial STM scores were in the low average range with substantial deficits in verbal and visuospatial working memory (Holmes, Gathercole, Place, Alloway, Elliott, & Hilton, 2009; Holmes, Gathercole, Alloway, *et al.*, 2010—see Figure 2). Of the total sample, 19.8% had impairments in verbal STM, which is close to the level of 16% that we would expect in the normal population. However, 38.6% had deficits in visuospatial STM, over half had impairments in verbal working memory (50.6%) and 63.9% had very poor visuospatial working memory.

It is possible that working memory problems may be the cause of the inattentive and distractible behavior associated with ADHD. To complete a task successfully, working memory resources support the maintenance of task goals as well as the different elements of the ongoing mental activity to achieve the goal—it enables us to stay on task and focus on the salient aspects of the task. Poor working memory function may therefore cause attention to shift away from the task at hand, resulting in the loss of part or all of the necessary information needed for task completion. This will

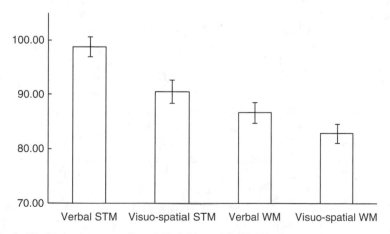

Fig. 2. Working memory profiles of 50 children with ADHD, from Holmes et al. (2010).

result in task failure, and as a consequence, individuals with ADHD may well appear to have short attention spans and to be distractible (Holmes, Gathercole, Alloway, *et al.*, 2010).

In summary, impairments in working memory are associated with a broad range of genetic and neurodevelopmental disorders of learning, with markedly distinct profiles of deficit characterizing different disorders. Deficits in the verbal domain are associated with specific language difficulties, such as SLI and dyslexia and are also characteristic of individuals with Down syndrome who experience severe language delays and difficulties. Conversely, children with William's syndrome exhibit domain-specific impairments in visuospatial memory. Unlike the other disorders discussed here, ADHD is associated with a domain-general impairment in working memory combined with deficits in visuospatial STM. There is now abundant evidence that tasks that require storage but no further processing of visuospatial material depend significantly on the domain-general resources of working memory (Miyake *et al.*, 2001; Wilson, Scott, & Power, 1987), rather than a distinct visuospatial store. This is particularly true in young children (Alloway *et al.*, 2006). Thus, the memory profile of children with ADHD corresponds to a singular impairment in the domain-general working memory system, which may cause the inattentive behavior that is characteristic of the disorder (Holmes, Gathercole, Alloway, *et al.*, 2010). Children with general reading difficulties have deficits in all aspects of working memory, whereas children with mathematical difficulties have severe impairments in verbal and visuospatial working memory and visuospatial STM, but not verbal STM. Children with poor working memory also have pervasive domain-general working memory deficits, akin to children with reading difficulties, mathematical difficulties, and ADHD. These are twinned with substantial impairments in visuospatial STM and poor verbal STM (see Figure 3).

E. COGNITIVE PROFILE—SUMMARY

To summarize, due to the key role working memory plays in supporting learning, both in classroom activities and in online mental activities, over 80% of children with small working memory capacities struggle in reading and mathematics (Gathercole & Alloway, 2008). Related to this, poor working memory function is characteristic of a number of developmental disorders of learning, including language, mathematical, and behavioral difficulties. Children with poor working memory also have deficits that are manifest in a variety of cognitively demanding activities, such as planning, sustained attention, problem solving, and IQ tasks. It is

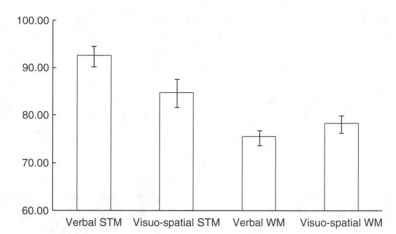

Fig. 3. Working memory profiles of 50 children with poor working memory, from Gathercole et al. (2010).

suggested, however, that they do not have selective impairments in high-level executive functions. Rather, limited memory resources constrain performance on a range of executive tasks (Gathercole *et al.*, 2010).

III. Social and Behavioral Profile

Poor working memory is associated with relatively normal social integration, self-esteem, and emotional control. However, high levels of inattentive and distractible behavior accompany working memory problems and individuals with poor working memory have difficulties maintaining focused behavior in practical situations.

It is now widely recognized that the majority of problems in individuals with poor working memory are related to inattentive and distractible behaviors. Both children and adults with low memory experience difficulties in practical situations that require maintained and focused attention. Kane, Brown, McVay, Silvia, Myin-Germeys, & Kwapil, (2007) found that typically developed adults with low working memory spans were more likely to "zone out" when engaged in demanding ongoing activities than individuals with higher working memory spans. They asked individuals to rate their behavior on several dimensions at eight random points during the day. Those with higher working memory spans were less likely to report instances of mind wandering and were able to maintain on task thoughts better during challenging cognitive tasks than those with poor working memory.

Poor working memory function is also closely associated with inattentive behavior in children. In a nonclinical sample, Aronen and colleagues found children with low working memory performance were reported by teachers to have more academic and attentional difficulties at school than children with good working memory performance (Aronen, Vuontela, Steenari, Salmi, & Carlson, 2005). Similarly, children identified solely on the basis of poor working memory skills have high levels of inattentive and distractible behavior. Teachers often describe them as having short attention spans and rarely say that they have memory problems (Gathercole, Alloway, *et al.*, 2006). Furthermore, when asked to rate behavior on commonly used checklists such as the Conner's Teacher Rating Scales (Conners, 1997), teachers typically judge children with poor working memory to be highly inattentive with high levels of distractibility. Over 70% of children aged 5 or 6 years with low working memory have markedly atypical scores on the cognitive problems/inattention subscale of the Conner's checklist (75% reported in Alloway *et al.*, 2009a, 2009b studies of 53 children; 79% reported in Gathercole, Alloway, *et al.*'s (2008) and Gathercole, Durling, *et al.*'s (2008) studies of 29 children). Figures for older children range from 58% (Alloway *et al.*, 2009a, 2009b) to 70% (Gathercole, Alloway, *et al.*, 2008). Gathercole, Alloway, *et al.* (2008) found that the majority of elevated scores were largely due to high ratings on problem behaviors that relate to inattention and short attention spans. In stark contrast, they found that none of the children in a comparison group of 20 children with typical working memory had atypically high levels of inattentive behavior.

ADHD in childhood is also characterized by both working memory deficits and inattentiveness (Holmes, Gathercole, Alloway, *et al.*, 2010; Klingberg *et al.*, 2005; Martinussen & Tannock, 2006; McInnes *et al.*, 2003; Willcutt, Doyle, *et al.*, 2005; Willcutt, Pennington, *et al.*, 2005). The co-occurrence of working memory and attentional problems in poor working memory and ADHD groups suggests there may be substantial overlap in the behavioral characteristics of the two groups. In a recent study, we directly compared teacher behavior ratings for 59 children with a diagnosis of ADHD and 27 children of the same age with low working memory (see Alloway, Gathercole, Holmes, Place, & Elliott, 2009). Teachers were asked to rate the extent to which a child has shown problem behaviors in school over the past month on the Conners' Teacher Rating Scale Revised Short-Form (Conners, 1997). Overall, teacher ratings of oppositional and hyperactive behaviors were significantly elevated in the ADHD group, while ratings of cognitive problems/inattention were elevated in both the ADHD and low working memory groups. As a consequence of high ratings on individual subscales, scores for both groups were also elevated on the ADHD index of the Conners' scale (Conners; see Figure 4). The inattentive symptoms observed in children with working memory deficits, which are also commonly associated with ADHD, most likely occur when overloaded working memory systems enable interference from irrelevant information to disrupt goal-directed behavior.

Beyond attentional problems, children with low working memory are typically reserved in group discussions in the classroom, but integrate well with friends and peers in less formal situations outside of the classroom (Gathercole, Alloway, *et al.*, 2008). Outgoing and humorous children with poor working memory rarely volunteer information in the classroom or raise their hand to answer questions, possibly because their poor memory skills make it hard for them to participate—teachers typically ask questions about recent activities which they may be unable to answer because they have forgotten the relevant information (Gathercole, Alloway, *et al.*, 2008).

Related to this, poor working memory function is not strongly associated with low self-esteem. Of 113 children with low memory ability, Alloway *et al.*, 2009a, 2009b found that overall levels of self-esteem were either at the good or vulnerable levels (43% and 39% of the sample, respectively). Only 12% scored at the very low end of the scale, which is characterized by those who may be depressed and need constant support

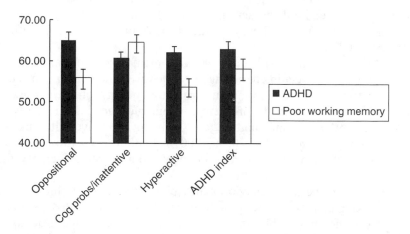

Fig. 4. Behavioral profiles of children with ADHD and children with poor working memory, from Holmes et al. (2010).

and encouragement (Morris, 2002). This demonstrates that very few children with poor working memory, who typically have poor academic success, have low self-esteem and is consistent with literature showing little association between global self-esteem and academic performance both in the general population (Baumeister, Campbell, Krueger, & Vohs, 2003; Marsh & Craven, 2006) and in those with learning difficulties (e.g., Snowling, Muter, & Carroll, 2007).

Emotional problems are not a hallmark characteristic of children with poor working memory, although studies that have examined teacher ratings report that approximately 50% of children identified as having poor working memory are also perceived to have problems with emotional control and regulation. Alloway *et al.* reported that 38% of their sample of 113 children had levels of emotional control problems that reached clinical significance (Alloway *et al.*, 2009a, 2009b). Likewise, Gathercole, Alloway, *et al.* (2008) reported that 45% of children aged 5/6 years with low working memory and 48% of children aged 9/10 years with low working memory obtained high ratings of problem behaviors relating to emotional control. It is possible that the incidence of emotional problems associated with poor working memory is a consequence of the number of children with poor working memory who have other comorbid disorders, such as ADHD or oppositional defiance disorder, which are more commonly associated with emotional and behavioral difficulties. Consistent with this view, children with low working memory have mildly elevated levels of oppositional and hyperactive behaviors in comparison to normative samples (Alloway *et al.*), and there is substantial overlap between the behavioral characteristics of children with low working memory and ADHD (e.g., Alloway *et al.*, 2009; Aronen *et al.*, 2005; Lui & Tannock, 2007).

Teachers of children with poor working memory rate them as having problem behaviors relating to a range of executive functions. In particular, they experience problems in monitoring the quality of their work, in generating new solutions to problems, planning/organizing written work, and large amounts of information, and in being proactive initiating new tasks (Alloway *et al.*, 2009, 2009a, 2009b; Gathercole, Alloway, *et al.*, 2008). As discussed earlier in this chapter, poor working memory may underpin this range of difficulties.

In summary, the key behavioral difficulties observed in children with poor working memory relate to inattention. Teachers view them as highly inattentive and distractible and judge them to have problem behaviors related to poor executive functioning. These behaviors are most likely the consequence of memory overload during complex and challenging mental activities, although further research is needed to test the direction of causality between poor attention, executive function problems, and working memory difficulties. In terms of social profiles, children with poor working memory are typically socially integrated, although they can be reserved in large group situations.

IV. Interventions

Three principal methods have been employed to reduce the difficulties that arise from poor working memory. One approach is to adapt the child's environment to minimize memory loads to facilitate classroom learning (e.g., Elliott, Alloway, Gathercole, Holmes, & Kirkwood, 2010; Gathercole & Alloway, 2008). The other two methods focus on improving working memory directly; one involves teaching children to use memory strategies to improve the efficiency of working memory (e.g., St. Clair-Thompson, Stevens, Hunt, & Bolder, 2010), the other involves directly training working memory through repeated practice on working memory tasks (e.g., Holmes, Gathercole, & Dunning, 2009; Holmes, Gathercole, Place, Dunning, Hilton, & Elliott, 2009; Klingberg *et al.*, 2005).

A. CLASSROOM INTERVENTION

This approach focuses on increasing teacher awareness of working memory problems and encouraging them to adapt their approach to teaching to reduce memory loads in the classroom. It also encourages teachers to help children with poor working memory to use strategies to overcome their cognitive weaknesses. Developed by Gathercole and colleagues, it is guided by seven key principles that are designed to decrease task failures, improve confidence, and accelerate rates of learning in children with low working memory.

The first stage of the intervention is to educate teachers about working memory and assist them in recognizing children who may be experiencing working memory failures. Teachers are often unaware that children with poor working memory have memory difficulties (Gathercole, Alloway, *et al.*, 2008). It is therefore paramount to teach them about the concept of working memory, to illustrate the contexts in which working memory plays a role in everyday classroom activities, and to emphasize the fact that working memory failures are often associated with inattentive behavior. This increased understanding can then be used to detect children who may have poor working memory (Gathercole & Alloway, 2008).

An integral part of supporting children with poor working memory in the classroom is to monitor how they cope with mentally challenging activities. This principle is closely related to detecting warning signs of memory overload, but goes further in that teachers and support assistants are encouraged to monitor whether children have forgotten crucial information during different activities. They can do this by asking simple questions such as "What were you going to write down?"

The third principle focuses on teachers evaluating learning activities to identify those that will be problematic for children with small working memory capacities. These include activities that place heavy demands on working memory, such as those that are overly long or include excessively long sequences, those that include large amounts of meaningless or unfamiliar material, and those that are complex and involve significant mental processing.

Directly related to this, the fourth aspect involves teachers developing and restructuring learning activities to reduce working memory loads. This can be done by reducing the amount of information a child has to remember, increasing the familiarity and meaningfulness of the material children are working with, and simplifying complex activities. So, for example, teachers might consider reducing the number of steps in an activity or breaking a long activity into several shorter activities, relating new material to previously acquired knowledge, simplifying linguistic structures, and reducing the length of sentences used to explain complex activities.

The fifth principle targets the profound difficulties children with poor working memory experience when trying to remember instructions and task information. It encourages teachers to frequently repeat important information, including classroom management instructions, task-specific instructions, and detailed information intrinsic to an activity. Teachers are encouraged to foster an environment in which children with poor working memory can ask for information to be repeated and to pair children with low working memory with those who do not struggle to remember information—this enables them to ask for information to be repeated in a less conspicuous way.

The final two principles involve teachers encouraging children to help themselves. First, teachers should provide and promote the use of memory aids such as wall charts and posters, lists of useful spellings and personalized dictionaries, counters, number lines, multiplication grids, calculators, memory cards, and audio recorders. Second, children with poor working memory should be encouraged to develop their own strategies to support their weak memory skills. These strategies include asking for help, rehearsing important information, note-taking, making links between new and previously learned information to activate support from long-term memory, and place-keeping and organizational strategies such as using flow-charts and diagrams.

Elliott and colleagues recently evaluated the effectiveness of this approach for boosting the learning outcomes of children with poor working memory (Elliott *et al.*, 2010). They found that the extent to which teachers implemented the principles of the working memory intervention predicted the children's literacy and mathematical skills and that teachers were enthusiastic about the ways in which their understanding and practice had improved as a result of the intervention. The long-term benefits for learning are not yet known, but this approach offers a practical starting point for teachers who are keen to help children with poor working memory.

B. STRATEGY TRAINING

The second approach to alleviating problems associated with poor working memory is to enhance working memory through training children to use strategies. Strategies are mentally effortful, goal-directed processes that improve the efficiency of working memory. These mechanisms include repeatedly rehearsing the to-be-remembered information aloud or in one's head, creating a sentence or story from the words, or generating visual images of the information. Individuals with high memory spans use strategies more than individuals with low spans (e.g., Engle, Cantor, & Carullo, 1992; Friedman & Miyake, 2004) and individual differences in strategy production account for individual differences in working memory task performance in adults (Dunlosky & Kane, 2007). Strategy use emerges during childhood, leading to the suggestion that developmental increases in working memory are at least partly mediated by their onset (Gathercole, 1999). For example, rehearsal does not emerge until about 7 years of age (e.g., Gathercole, 1998), with other strategies, such as organization and grouping (e.g., Bjorkland & Douglas, 1997) and chunking (e.g., Ottem, Lian, & Karlsen, 2007) developing later. Although strategy use is not spontaneous in young children, they will attempt to employ them when explicit instructions are given (e.g., Ornstein, Baker-Ward, & Naus, 1988; Ornstein & Naus, 1985). This has led to the suggestion that training children to employ memory strategies could facilitate their working memory functioning (e.g., St. Clair-Thompson & Holmes, 2008).

Training adults to use strategies facilitates their short-term and working memory abilities. Improvements in STM tasks have been reported when participants engage in strategies, such as rehearsal (e.g., Broadley, MacDonald, & Buckley, 1994; Gardiner, Gawlick, & Richardson-Klavehn, 1994; Rodriguez & Sadoski, 2000), visual imagery (e.g., De La Iglesia, Buceta, & Campos, 2005), grouping or chunking information (e.g., Black & Rollins, 1982; Carr & Schneider, 1991; Lange & Pierce, 1992), or creating stories from the to-be-remembered information (e.g., McNamara & Scott, 2001). Rehearsal, story generation, and visual imagery strategies also improve performance on working memory span tasks (Turley-Ames & Whitfield, 2003; McNamara & Scott). These gains, however, often do not extend beyond trained tasks and are not sustained, meaning they are unlikely to yield substantial benefits for the multiplicity of situations in which we depend on working memory.

Recently, St. Clair-Thompson and colleagues have shown that providing memory strategy training significantly improves working memory in children (St. Clair-Thompson & Holmes, 2008; St. Clair-Thompson *et al.*, 2010). In their first two studies, they measured the impact of strategy training on short-term and working memory in small groups of 6 and 7 and 7 and 8 year olds (St. Clair-Thompson & Holmes, 2008). The strategy training was provided in the form of a computerized adventure game called Memory Booster (Leedale, Singleton, & Thomas, 2004), which taught and encouraged the use of rehearsal, visual imagery, story generation, and grouping. Children completed two training sessions per week, each lasting approximately 30 minutes, for 6–8 weeks. In the first study, children were assessed on measures of verbal and visuospatial STM and verbal working memory both before and after training. There were no significant gains in STM for either age group. There were, however, significant gains in verbal working memory for the younger group of children (6 and 7 year olds). This pattern of improvement was reflected in a second study, in which a measure of children's ability to remember and follow instructions was also included. Again, there was no impact of training on STM for either age group, but both verbal working memory and performance on the following instructions task improved in the younger children (St. Clair-Thompson & Holmes).

Together, these initial studies demonstrated that working memory could be improved in young children through strategy training and there was preliminary evidence that gains generalized to a working memory-loaded classroom activity. There were no benefits of strategy training for older children, who were aged 7 and 8 years. It is possible that these children were already using efficient memory strategies meaning strategy training was unlikely to benefit their memory performance.

In a more recent study, St. Clair-Thompson and her colleagues investigated the potential usefulness of memory strategy training further

(St. Clair-Thompson *et al.*, 2010). A total of 254 children aged 5–8 years participated in the study. Half of the children completed Memory Booster strategy training and half received no intervention. All children were tested on measures of verbal and visuospatial STM and verbal working memory before and after training and subgroups completed a following instructions task, a mental arithmetic task and standardized assessments in reading, arithmetic, and mathematics. This time, significant training gains were found in verbal short-term and working memory, but not in visuospatial STM. Significant improvements were also reported in mental arithmetic and following instructions, but no improvements were found in standardized ability measures either immediately after training or 5 months later. The same patterns of gains were observed both in children aged 5/6 and 7/8 years, and also in children with average working memory and poor working memory (in this case classed as standard scores below 70; see St. Clair-Thompson *et al.* for further details).

This recent work on memory strategy training in children suggests it could provide a means of improving working memory in children. However, the transfer of benefits to other tasks seems limited. Although there is some evidence of transfer to tasks that allow for the use of memory strategies, such as following instructions, there is no evidence that gains transfer to standardized ability measures.

C. WORKING MEMORY TRAINING

The final approach to remediating poor working memory function is to train it through repeated practice on working memory tasks. Studies that attempted to improve working memory using this method in the 1970s and 1980s only reported moderate training gains, which were in the form of faster reaction times, not increases in working memory capacities *per se*, and there was no evidence that gains were transferable to nontrained working memory tasks or to other cognitive measures (Kristofferson, 1972; Phillips & Nettlebeck, 1984). Other training studies, such as those conducted by Hulme and Muir (1985), have demonstrated that training processes crucial to efficient processing in working memory such as articulation and rehearsal rate improve memory span, although only very slightly. More recent evidence indicates that intensive adaptive working memory training specifically on tasks that tax working memory may lead to more dramatic gains on trained and nontrained working memory tasks and also on other cognitive measures.

Jaeggi and colleagues found improvements on trained working memory tasks that transferred to marked increases in performance on a general

IQ test in healthy adults, using a paradigm known as *n*-back. This involves the continuous presentation of sequence of stimuli. Participants are required to indicate whether the current stimulus matches the one that was presented *n* steps earlier in the sequence. The task can be adjusted by increasing or decreasing the number of steps to make the task more or less difficult. Jaeggi and colleagues used an extremely cognitively demanding version of this task in which participants were presented with two rows of stimuli simultaneously. The top row showed individual spatial locations marked in a box, the bottom showed a string of individual letters. Participants were required to decide for each string whether the current stimuli matched the one presented *n* items back in the series. The number of items back varied according to participants' performance, increasing by one if performance improved and decreasing by one if performance dropped (Jaeggi, Buschkuel, Jonides, & Perrig, 2008).

Five groups of participants were included. Four groups completed between 8 and 19 training sessions and a fifth group was not trained. IQ was assessed pre- and post-training. Overall, there was a significant training effect, with all four training groups improving on the working memory tasks. Importantly, there was a transfer effect to performance on the general IQ test. This was strictly training related and could not be attributed to individual differences in pre-training working memory or IQ scores, gains in working memory as measured by changes on complex span tasks, or to the training itself. The gains in general IQ were dose dependent, with more training equating to bigger gains.

Jaeggi *et al.* (2008) suggest the transfer effect may result from the shared features between the training tasks and IQ measure, such as attentional control and the requirement to hold multiple goals in mind. However, they also propose that changes in performance on the IQ test may result from improvements in skills that were not working memory capacity related but that were also trained by the dual *n*-back task, for example, multiple-task management skills. It should be noted, however, that other researchers have proposed these findings do not represent a true transfer effect; rather they are a consequence of testing procedure used to administer the IQ test (see Moody, 2009).

Using a similar but less cognitively challenging program, Jaeggi and colleagues have also demonstrated that working memory can be improved through training in older adults (Buschkeuhl *et al.*, 2008). In this study, 13 women aged 80+ years trained on three working memory tasks and two speed of processing tasks two times a week for 3 months. The working memory tasks included recalling sequences of colors or animals in the correct order. Some of the tasks included an explicit processing element, which, for example, required participants to decide if the animals were in the correct orientation prior to recalling the sequence. The processing speed tasks were forced reaction time tasks in which participants had to make a decision and respond as quickly as possible to stimuli presented on the screen. A control group of 19 women aged 80+ years formed an active control group who completed a physical intervention, training on a bicycle ergo meter for the same amount time as the memory training group. Working memory and episodic memory were assessed before, immediately after and 1 year after training for both groups.

Performance on the trained tasks improved significantly for those in the memory training condition. There was evidence of transfer to nontrained visual working memory tasks immediately after training, although these transfer effects did not extend to a digit span task or measures of episodic memory. There were no significant differences between the memory and physical training groups 1 year after training. These results suggest it is possible to improve visual working memory in old–old adults through training in the short-term, but that it is unlikely these gains will be sustained. Furthermore, due to the complex nature of the training, which included both speed of processing and working memory training, it is difficult to determine the source of the gains in visual working memory.

An alternative method that has also produced positive results is the Cogmed Working Memory Training Program (CWMT), which was developed by Klingberg and colleagues at the Karolinska Institute in Sweden. In this program, individuals train intensively over several weeks on adaptive working memory tasks. The training tasks require the immediate serial recall of either verbal or visuospatial information, with some of the tasks requiring explicit processing prior to recall. Participants train for 20–25 days, each day completing 8 different tasks selected from a bank of 13, which takes approximately 30–45 minutes. The tasks vary across training days to maintain interest and positive verbal and visual feedback is given on some trials. The difficulty of the training tasks is automatically adjusted on a trial-by-trial basis to match the participant's current working memory capacity, which maximizes the training benefits.

The Cogmed program is perhaps the most widely validated memory training program, with studies showing dramatic improvements in working memory following training in both children and adults. Crucially, it is the only program to date that has been used to directly train working memory in children. In the very first trials, Klingberg's team used an early form of the training program that included only four training tasks: a Corsi block-like visuospatial memory task; two verbal tasks—backward digit and letter span; and a choice reaction time task. In a double-blind, placebo-controlled study, this primitive form of intensive and adaptive working memory training significantly improved performance on non-trained STM tasks, digit recall and Corsi block, and a test of nonverbal reasoning in a small sample of children with ADHD (treatment group $n=7$, control group $n=7$). Motor activity, as measured by the number of head movements during a computerized test, was significantly reduced in the treatment group, and performance on a response inhibition task also significantly improved following training. There were no significant changes in performance for the control group, who completed a placebo version of the program in which the difficulty of the training tasks was set a low level throughout the training period (span of 2 or 3 items for each task). In a second experiment, they used the same adaptive training program with four healthy adults. Significant improvements in performance were reported both on the trained tasks and on a nontrained visuospatial memory task, a Stroop task and a nonverbal reasoning task (Klingberg, Forssberg, & Westerberg, 2002).

Klingberg's team later extended their work to evaluate the effects of training in a larger group of children with ADHD ($n=53$) in a randomized controlled trial. The intensity of the treatment training program was increased to 90 trials per day on three working memory tasks (remembering the position of objects in a 4×4 grid or remembering phonemes, letters, or digits) for 20–25 days. As before, the placebo version included an identical set of tasks to the treatment program, which were set at a low level throughout training. Children were randomly assigned to each condition, with 27 completing the adaptive treatment program and 26 completing the placebo version. Overall, the treatment group improved significantly more than the comparison group on a non-trained measure of visuospatial STM. These effects persisted 3 months after training. In addition, significant treatment effects were observed in response inhibition, complex reasoning, and verbal STM, and there were significant reductions in parent ratings of inattention and hyperactivity/impulsivity following training. Reductions in ratings of cognitive problems following training were also reported in a pilot study with 18 adults more than 1 year after a stroke. As before, there were significant improvements in trained and nontrained working memory tasks and there was also a significant decrease in the patients' self-ratings of cognitive problems in daily life (Westerberg *et al.*, 2007).

The team at the Karolinska Institute has now extended their work on memory training to preschool children (Thorell, Lindqvist, Bergman, Bohlin, & Klingberg, 2009). On the basis of the strong theoretical connection between inhibition and working memory (see Engle & Kane, 2004; Roberts & Pennington, 1996), the overlapping areas of neural activation during working memory and inhibition tasks (McNab *et al.*, 2008) and the transfer of training effects to the Stroop task in their early studies, they decided to compare the effects of visuospatial working memory training and inhibition training in very young children.

Four groups of preschool children aged 4 and 5 years were included in the study. One group completed visuospatial working memory training ($n=17$). A second group completed inhibition training ($n=18$), the third ($n=14$) completed a placebo version of the memory training, as per previous studies, and a fourth group formed a passive control group ($n=16$). Those in the training groups completed adaptive training of either visuospatial working memory or inhibition for 15 minutes a day every day that they attended preschool over a 5-week period. Each day they completed three of five possible tasks, which rotated across the training period to maintain interest. The five visuospatial working memory training tasks required children to recall sequences of nonverbal information in the correct order. The inhibition training consisted of five tasks that mirror well-known inhibition paradigms—two go/no-go tasks and two stop signal tasks designed to train response inhibition and a flanker task designed to train interference control. Outcome measures for all groups included nontrained measures of interference control, response inhibition, forward and backward Corsi block, forward and backward digit recall, sustained attention, and problem solving.

Children in the working memory training group improved significantly on all trained tasks, whilst those in the inhibition training group improved only on the trained go/no-go and interference control tasks. Working memory training led to significant gains in both nontrained verbal and visuospatial memory tasks and attention, but there was no significant transfer to performance on nontrained inhibition tasks. There was no significant change in performance on nontrained tasks for children in the inhibition training, placebo, or passive control groups. Overall, the data from this study show that working memory can be trained in typically developing children as young as 4 years and, perhaps most importantly, demonstrates that different cognitive functions vary in how easily they can be modified by intensive practice (Thorell *et al.*, 2009). The results of this study point to generalized benefits of training working memory, but of limited effects of training other executive functions such as inhibition.

On the basis of the early success of working memory training with children, we have recently conducted our own independent evaluations of the CWMT program. Like Klingberg, our first study was with children with ADHD, who we know have significant and substantial deficits in working memory (Holmes, Gathercole, Alloway, *et al.*, 2010; Holmes, Gathercole, Place *et al.*, 2010; Martinussen *et al.*, 2006).

The primary treatment option for reducing the behavioral symptoms of ADHD is psycho stimulant medication in the form of methylphenidate or amphetamine compounds, which also enhance visuospatial working memory (Bedard, Jain, Hogg-Johnson, & Tannock, 2007). The aim of our first study was to therefore compare the impacts of working memory training and psycho stimulant medication on the separate subcomponents of working memory. We recruited 25 children aged 8–11 years with a clinical diagnosis of combined type ADHD, who were receiving quick release medication for their ADHD symptoms. All children completed assessments of verbal and visuospatial STM and working memory both before and after training and on and off medication. The training paradigm consisted of 20–25 sessions on the adaptive program developed by Cogmed, which this time consisted of 10 different working memory tasks. Children trained on 8 of the 10 tasks every day, completing 115 trials per session.

Both interventions had a significant impact on children's working memory, but differential patterns of change were associated with each approach. While medication led to selective improvements in visuospatial working memory, training led to improvements in all aspects of working memory. Crucially, these gains were sustained 6 months after training ceased (see Figure 5).

Children's IQ was not affected by either intervention. The impact of medication on nonverbal aspects of working memory only most likely reflects the predominant influence of medication on right hemisphere brain structures that are associated with visuospatial working memory (e.g., Bedard *et al.*, 2007). The generalized impact of working memory training in this group may have very practical benefits for learning in children with ADHD. Although medication helps to control the adverse behavioral symptoms of the disorder, providing improved working memory resources through working memory training holds the promise of providing improved support for learning for in this group (Holmes, Gathercole, Place, Dunning, *et al.*, 2009).

We also have encouraging preliminary data to suggest that working memory training benefits other groups of children. In a recent study, we trained 25 typically developing children aged 9/10 years in a large group in school and found significant improvements in the aspects of working memory that are most strongly associated with learning, namely visuospatial STM and verbal and visuospatial working memory (Holmes, Dunning, & Gathercole, 2010a; see Figure 6).

Furthermore, we have some indication that memory training will benefit children with dyslexia (Holmes, Dunning, & Gathercole, 2010b). Figure 7 summarizes data showing the impact of working memory training on the

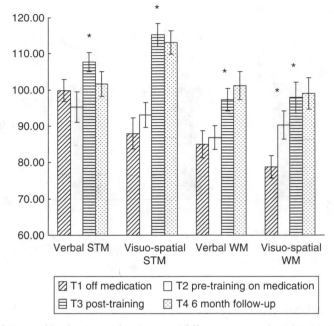

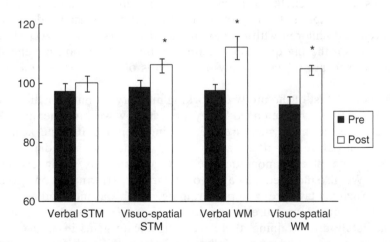

Fig. 5. *Impact of medication and training on different aspects of working memory, from Holmes, Gathercole, Place, Dunning, et al., 2009.* Note. *Asterisk above a bar denotes a significant change in scores from the previous testing point.*

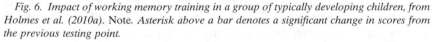

Fig. 6. *Impact of working memory training in a group of typically developing children, from Holmes et al. (2010a).* Note. *Asterisk above a bar denotes a significant change in scores from the previous testing point.*

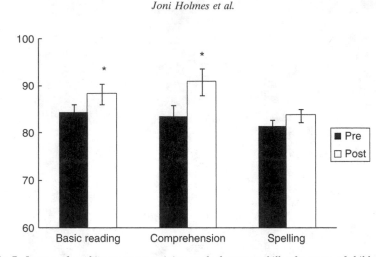

Fig. 7. *Impact of working memory training on the language skills of a group of children with dyslexia, from Holmes et al. (2010b). Note. Asterisk above a bar denotes a significant change in scores from the previous testing point.*

reading skills of a group of 19 children aged 8–11 years with dyslexia. In this study, we found significant improvements in both single word reading and reading comprehension following adaptive training, with reading comprehension scores improving from the deficit to average range. There was no significant improvement in spelling post-training.

Thus far, the benefits of focused working memory training have been discussed in terms of improvements in adults, typically developing children and children with developmental disorders such as ADHD and dyslexia. So, the big question remains—is this intervention beneficial to children who are selected solely on the basis of poor working memory? The answer is yes.

In a direct test of the utility of working memory training for this group, we identified 42 children aged 9/10 years with low working memory skills and assessed them on measures of working memory, IQ, and academic attainment before and after training either on the standard adaptive program or the placebo non-adaptive program used by Klingberg *et al.* (2005). We also included a measure of children's performance on a following instructions task pre- and post-training, which was included as a more practically based measure of working memory use in the classroom. Immediately after training, there were significant gains in all aspects of working memory and on the following instructions task for children in the adaptive condition ($n=20$). In all cases, the gains were significantly greater for the adaptive than the nonadaptive group ($n=22$). There was no immediate impact of training on standardized reading and mathematics tests or on IQ. Follow-up assessments revealed the training gains were sustained in the adaptive group 6 months after training, at which point performance on the standardized mathematics test also improved significantly. These results provide the first solid evidence that commonplace deficits in working memory and associated learning difficulties can be ameliorated, and possibly even overcome, by intensive adaptive training over a relatively short period (Holmes, Gathercole, & Dunning, 2009).

So, what is driving the changes in working memory following training? One possibility is that intensive training induces long-term plasticity in the brain regions that serve working memory. In two neuroimaging studies, Olesen, Westerberg, and Klingberg (2004) showed increased activation in the parietal and prefrontal cortices following memory training. In their first study, they reported an increase in brain activity in both of these regions in three subjects following training on both verbal and visuospatial working memory tasks. In the second study, eight adults were scanned five times during training on three visuospatial working memory tasks over a 5-week period. Again, increases in neuronal activity were observed in the prefrontal and parietal regions. In a single-subject analysis, Westerberg and Klingberg (2007) showed training-induced changes were not due to activations of any additional area that was not activated before training. Rather, they observed that areas where task-related activity was seen increased in size following training. Related to this, changes in the density of prefrontal and parietal cortical dopamine receptors have been reported after memory training. Either too much or too little stimulation of D1 receptors results in impaired working memory task performance (Cai & Arsten, 1997; Lidow, Williams, & Goldman-Rakic, 1998; Vijayraghavan, Wang, Birnbaum, Williams, & Arnsten, 2007; Williams & Goldman-Rakic, 1995). Thus, training-induced decreases in the binding potential of the receptor D1 associated with increases in working memory capacity were interpreted as demonstrating a high level of plasticity of the D1 receptor system (McNab *et al.*, 2009).

Overall, Klingberg's team has shown that training induces changes in two brain regions, the parietal and prefrontal cortices, which are both associated with working memory. Prefrontal activation is positively correlated with children's working memory capacity (Klingberg *et al.*, 2002; Kwon, Reiss, & Menon, 2002) and fronto-parietal networks are related to success on working memory tasks (Pessoa, Gutierrez, Bandettini, & Ungerleider, 2002; Rypma & D'Esposito, 2003). Their work suggests cortical and biochemical changes result from practice on working memory tasks.

The working memory training program may also promote self-awareness and the development of compensatory strategies to overcome areas of weakness. Introspective reports from children in our own training studies support this notion. When asked what they thought had helped them to improve on the training activities, 27% of children with low working memory and 27% of children with ADHD reported focusing more on the presented information or concentrating harder. A further 37% of children with low working memory and 67% of children with ADHD reported using a variety of other strategies, including rehearsal or tracing the patterns on the computer screen with their eyes (Holmes, Gathercole, & Dunning, 2009; Holmes, Gathercole, Place, Dunning, *et al.*, 2009). These verbal reports support the idea that training may enhance attentional focus and stimulate the development of a range of strategies that capitalize on relative cognitive strengths.

The data presented in this section demonstrate that it is possible to train working memory through repeated practice on working memory tasks. Importantly, they provide an early indication that the learning difficulties that so often accompany working memory deficits can be ameliorated, to some extent, by training. This finding is highly relevant to professionals looking for effective interventions for children with poor working memory and also for children with a wide range of disorders that are associated with both low working memory and poor academic performance. Although we have some insight into the nature of the cognitive and neural changes that underpin the gains in working memory, further research is needed in this area.

D. INTERVENTIONS—SUMMARY

The difficulties faced by children with low working memory can be targeted in different ways: (i) by adjusting the child's environment to reduce working memory loads, (ii) by training and promoting the use of effective strategies to improve the efficiency of working memory, and (iii) by training working memory directly using an adaptive, computerized program. The latter approach appears to be the most direct, useful, and well-validated method for boosting both poor working memory and learning in children. Strategy training has also been shown to boost working memory, both in children with low and average working memory, but the evidence for transfer effects to improved academic outcomes is far less persuasive. The classroom-based intervention, which focuses on changes in the ways children with poor working memory are taught, is perhaps the most practical approach as it builds upon and promotes good teaching practice. Although the extent to which teachers apply the principles of the intervention correlates with attainment, the benefit of this method for improving learning outcomes is as yet somewhat unclear. All three approaches require further investigation to determine the long-term effects on working memory and learning.

V. Summary and Future Directions

In conclusion, poor working memory affects approximately 15% of children. It is characterized by inattentive, distractible behavior that is accompanied by failures to complete everyday activities that require focused or sustained attention. Typically, children with poor working memory have normal social integration and normal levels of emotional control and self-esteem. They may, however, appear reserved in large group situations. Over 80% of children with low working memory struggle in reading and mathematics and it has been suggested that they are likely to be those children who make poor academic progress, but who fall below the radar of recognition for SEN.

Beyond poor learning, children with poor working memory have a range of other cognitive problems that extend to low IQ and deficits in other executive functions including monitoring and planning, problem solving, and sustained attention. Although the direction of causality is uncertain, it is possible that limited working memory resources underpin this wide range of deficits.

There are three main approaches to alleviating the difficulties faced by children with poor working memory: a classroom intervention, strategy training, or direct working memory training. Of these, direct training is the most widely used and successful method. Dramatic gains in working memory have been reported following training in typically developing children and adults, adults following a stroke, children with developmental disorders such as ADHD and, most pertinent to this chapter, children with poor working memory. Importantly, the memory gains in children with poor working memory sustain over a 6-month period, at which point they transfer to gains in learning. This finding holds the promise for improving the long-term educational outcomes of children with poor working memory.

Although we now have a fairly comprehensive understanding of the social, behavioral, and cognitive profiles of children with poor working memory, there are a number of challenges for the future. The first is to investigate the long-term consequences of poor working memory. Do working memory problems persist into adolescence and adulthood? If so, how do they impact on everyday functioning? Do children with poor working memory go on to develop compensatory strategies as they age? The second challenge is to investigate further the future impacts of different interventions. For example, is there a long-range benefit to learning from the classroom intervention? Do the dramatic training-induced gains in working memory endure? Related to this, the third challenge lies in determining what is driving the changes in working memory following direct training. Finding that working memory can be improved through intensive practice poses a challenge to theories that suggest working memory has a relatively fixed capacity (e.g., Engle, Cantor, & Carullo, 1992), is highly heritable (Kremen *et al.*, 2007), and is relatively impervious to substantial differences in environmental experience and opportunity (Campbell *et al.*, 1997; Engel *et al.*, 2008). Although there is evidence that training may induce neural plasticity in the brain regions supporting working memory, or the development of compensatory strategies, further neuroimaging, cognitive, and behavioral research is needed to fully understand what underpins the changes in working memory performance.

REFERENCES

Alloway, T. P. (2007). *Automated working memory assessment.* Oxford, UK: Pearson Assessment.

Alloway, T. P., & Gathercole, S. E. (2006). *Working memory and neurodevelopmental conditions.* Hove, England: Psychology Press.

Alloway, T. P., Gathercole, S. E., Adams, A.-M., & Willis, C. S. (2005). Working memory abilities in children with special educational needs. *Educational and Child Psychology*, 22, 56–67.

Alloway, T. P., Gathercole, S. E., Holmes, J., Place, M., & Elliott, J. (2009a). The diagnostic utility of behavioral checklists in identifying children with ADHD and children with working memory deficits. *Child Psychiatry & Human Development*, 40, 353–366.

Alloway, T. P., Gathercole, S. E., & Kirkwood, H. (2008). *Working Memory Rating Scale.* Oxford, UK: Pearson Assessment.

Alloway, T. P., Gathercole, S. E., Kirkwood, H., & Elliott, J. (2009b). The cognitive and behavioral characteristics of children with low working memory. *Child Development*, 80, 606–621.

Alloway, T. P., Gathercole, S. E., & Pickering, S. J. (2006). Verbal and visuo-spatial short-term and working memory in children: Are they separable? *Child Development*, 77, 1698–1716.

Anderson, J. R., & Lebiere, C. (1998). *Atomic components of thought.* Hillsdale, NJ: Lawrence Erlbaum Associates, Inc.

Archibald, L. M., & Gathercole, S. E. (2006). Short-term and working memory in Specific Language Impairment. *International Journal of Language and Communication Disorders*, 41, 675–693.

Archibald, L. M., & Gathercole, S. E. (2007). The complexities of complex span: Storage and processing deficits in Specific Language Impairment. *Journal of Memory & Language, 57,* 177–194.

Aronen, E. T., Vuontela, V., Steenari, M. R., Salmi, J., & Carlson, S. (2005). Working memory, psychiatric symptoms, and academic performance at school. *Neurobiology of Learning and Memory*, *83*, 33–42.

Baddeley, A. D. (1996). Exploring the central executive. *Quarterly Journal of Experimental Psychology*, *49A*, 5–28.

Baddeley, A. D. (2000). The episodic buffer: A new component of working memory? *Trends in Cognitive Sciences*, *4*, 417–423.

Baddeley, A. D., Emslie, H., Kolodny, J., & Duncan, J. (1998). Random generation and the executive control of working memory. *Quarterly Journal of Experimental Psychology*, *51A*, 819–852.

Baddeley, A. D., & Hitch, G. J. (1974). Working memory. In G. H. Bower (Ed.), *The psychology of learning and motivation, 8*. New York: Academic Press.

Barnett, R., Maruff, P., Vance, A., Luk, E. S. L., Costin, J., Wood, C., *et al.* (2001). Abnormal executive function in attention deficit hyperactivity disorder: The effect of stimulant medication on age and spatial working memory. *Psychological Medicine*, *31*, 1107–1115.

Barrouillet, P., Bernardin, S., & Camos, V. (2004). Time constraints and resource sharing in adults' working memory spans. *Journal of Experimental Psychology: General*, *133*, 83–100.

Baumeister, R. F., Campbell, J. D., Krueger, J. L., & Vohs, K. D. (2003). Does high self-esteem cause better performance, interpersonal success, happiness, or healthier lifestyles? *Psychological Science in the Public Interest*, *4*, 1–44.

Bedard, A., Jain, U., Hogg-Johnson, S., & Tannock, R. (2007). Effects of methylphenidate on working memory components: Influence of measurement. *Journal of Child Psychology and Psychiatry*, *48*, 872–880.

Bjorkland, D. F., & Douglas, R. N. (1997). The development of memory strategies. In N. Cowan & C. Hulme (Eds.), *The development of memory in childhood* (pp. 201–246). Sussex: Psychology Press.

Black, M. M., & Rollins, H. A. (1982). The effects of instructional variables on young children's organisation and free recall. *Journal of Experimental Child Psychology*, *33*, 1–19.

Broadley, I., MacDonald, J., & Buckley, S. (1994). Are children with Down's syndrome able to maintain skills learned from short-term memory training programme? *Down's Syndrome: Research and Practice*, *2*, 106–112.

Brooks-Gunn, J., Klebanov, P. K., & Duncan, G. J. (1996). Ethnic differences in children's intelligence test scores: Role of economic deprivation, home environment, and maternal characteristics. *Child Development*, *67*, 396–408.

Bull, R., & Scerif, G. (2001). Executive functioning as a predictor of children's mathematical ability: Inhibition, switching and working memory. *Developmental Neuropsychology*, *19*, 273–293.

Buschkeuhl, M., Jaeggi, S., Hutchison, S., Perrig-Chiello, P., Dapp, C., Muller, M., *et al.* (2008). Impact of working memory training on memory performance in old–old adults. *Psychology and Aging*, *23*, 743–753.

Cai, J. X., & Arsten, A. F. T. (1997). Dose-dependent effects of the dopamine D1 receptor agonists A77636 or SKF81297 on spatial working memory in aged monkeys. *Journal of Pharmacology and Experimental Therapeutics*, *23*, 183–189.

Cain, K., Oakhill, J., & Bryant, P. (2004). Children's reading comprehension ability: Concurrent prediction by working memory, verbal ability, and component skills. *Journal of Educational Psychology*, *96*, 31–42.

Camos, V. (2008). Low working memory capacity impedes both efficiency and learning of number transcoding in children. *Journal of Experimental Child Psychology*, *99*, 37–57.

Campbell, T., Dollaghan, C., Needleman, H., & Janosky, J. (1997). Reducing bias in language assessment: Processing dependent measures. *Journal of Speech, Language and Hearing Research*, *40*, 519–525.

Carpenter, P. A., Just, M. A., & Shell, P. (1990). What one intelligence test measures: A theoretical account of the processing in the Raven Matrices test. *Psychological Review*, *97*, 404–431.

Carr, M., & Schneider, W. (1991). Long-term maintenance of organisational strategies in kindergarten children. *Contemporary Educational Psychology*, *16*, 61–72.

Chiappe, P., Hasher, L., & Siegel, L. S. (2000). Working memory, inhibitory control, and reading disability. *Memory & Cognition*, *28*, 8–17.

Conners, K. (1997). *Conners' teacher rating scale revised short-form*. New York: Multi-Health Systems Inc.

Conway, A. R. A., Jarrold, C., Kane, M. J., & Towse, J. N. (Eds.), (2007). *Variation in working memory*. New York: Oxford University Press.

Cowan, N. (2005). *Working memory capacity*. New York: Psychology Press.

Daneman, M., & Carpenter, P. A. (1980). Individual differences in working memory and reading. *Journal of Verbal learning and Verbal Behaviour*, *19*, 450–466.

Daneman, M., & Merickle, P. M. (1996). Working memory and language comprehension: A meta-analysis. *Psychonomic Bulletin and Review*, *3*, 422–433.

de Jong, P. F. (1998). Working memory deficits of reading disabled children. *Journal of Experimental Child Psychology*, *70*, 75–96.

de Jonge, P., & de Jong, P. F. (1996). Working memory, intelligence and reading ability in children. *Personality and Individual Differences*, *21*, 1007–1020.

De La Iglesia, J. C. F., Buceta, M. J., & Campos, A. (2005). Prose learning in children and adults with Down syndrome: The use of visual and mental image strategies to improve recall. *Journal of Intellectual and Developmental Disability*, *30*, 199–206.

Dunlosky, J., & Kane, M. J. (2007). The contributions of strategy use to working memory span: A comparison of strategy assessment methods. *Quarterly Journal of Experimental Psychology*, *60*, 1227–1245.

Edwards, J., & Lahey, M. (1998). Nonword repetitions of children with specific language impairment: Exploration of some explanations for their inaccuracies. *Applied Psycholinguistics*, *19*, 279–309.

Elliott, J. G., Gathercole, S. E., Alloway, T.P., Holmes, J., & Kirkwood, H. (2010). An evaluation of a classroom-based intervention to help overcome working memory difficulties and improve long-term academic achievement. *Journal of Experimental Child Psychology*, in press.

Ellis Weismer, S., Evans, J., & Hesketh, L. (1999). An examination of working memory capacity in children with specific language impairment. *Journal of Speech, Language, and Hearing Research*, *42*, 1249–1260.

Ellis Weismer, S., Tomblin, J. B., Zhang, X. Y., Buckwalter, P., Chynoweth, J. G., & Jones, M. (2000). Nonword repetition performance in school-age children with and without language impairment. *Journal of Speech, Language, and Hearing Research*, *43*, 865–878.

Engel, P. M. J., Santos, F., & Gathercole, S. E. (2008). Are working memory measures free of socio-economic influence? *Journal of Speech, Language and Hearing Research*, *51*, 1580–1587.

Engle, R. W., Cantor, J., & Carullo, J. J. (1992). Individual differences in working memory and comprehension: A test of four hypotheses. *Journal of Experimental Psychology: Learning, Memory, and Cognition*, *18*, 972–992.

Engle, R. W., Carullo, J. J., & Collins, K. W. (1991). Individual differences in working memory for comprehension and following directions. *Journal of Educational Research, 84,* 253–262.

Engle, R. W., & Kane, M. J. (2004). Executive attention, working memory capacity, and a two-factor theory of cognitive control. In B. H. Ross (Ed.), *The psychology of learning and motivation: Advances in research and theory* (Vol. 44, pp. 145–199). New York: Elsevier Science.

Engle, R. W., Kane, J. M., & Tuholski, S. W. (1999). Individual differences in working memory capacity and what they tell us about controlled attention, general fluid intelligence and functions of the prefrontal cortex. In A. Miyake & P. Shah (Eds.), *Models of working memory* (pp. 102–135). Cambridge: Cambridge University Press.

Friedman, N. P., & Miyake, A. (2004). The reading span test and its predictive power for reading comprehension ability. *Journal of Memory and Language, 51,* 136–158.

Gardiner, J. M., Gawlick, B., & Richardson-Klavehn, A. (1994). Maintenance rehearsal affects knowing not remembering; elaborative rehearsal effects remembering not knowing. *Psychonomic Bulletin and Review, 1,* 107–110.

Gathercole, S. E. (1998). The development of memory. *Journal of Child Psychology and Psychiatry, 39,* 3–27.

Gathercole, S. E. (1999). Cognitive approaches to the development of short-term memory. *Trends in Cognitive Sciences, 3,* 410–419.

Gathercole, S. E. (2010). *Working memory deficits in slow learners: Risk factor or cognitive phenotype?* Proceedings of the 8th Sepex conference, 1st Joint conference of the EPS and SEPEX. Granada, Spain: Sepex.

Gathercole, S. E., & Alloway, T. P. (2008). *Working memory and learning: A practical guide for teachers.* London, UK: Sage Publishing.

Gathercole, S. E., Alloway, T. P., Kirkwood, H. J., Elliott, J. G., Holmes, J., & Hilton, K. A. (2008). Attentional and executive function behaviors of children with poor working memory. *Learning and Individual Differences, 18,* 214–223.

Gathercole, S. E., Alloway, T. P., Willis, C. S., & Adams, A. M. (2006). Working memory in children with reading disabilities. *Journal of Experimental Child Psychology, 93,* 265–281.

Gathercole, S., & Baddeley, A. (1990). Phonological memory deficits in language disordered children: Is there a causal connection? *Journal of Memory and Language, 29,* 336–360.

Gathercole, S. E., & Baddeley, A. D. (1993). Phonological working memory: A critical building block for reading development and vocabulary acquisition? *European Journal of the Psychology of Education, 8,* 259–272.

Gathercole, S. E., Brown, L., & Pickering, S. J. (2003). Working memory assessments at school entry as longitudinal predictors of National Curriculum attainment levels. *Educational and Child Psychology, 20,* 109–122.

Gathercole, S. E., Durling, M., Evans, S., Jeffcock, E., & Stone, S. (2008). Working memory abilities and children's performance in laboratory analogues of classroom activities. *Applied Cognitive Psychology, 22,* 1019–1037.

Gathercole, S. E., & Hitch, G. J. (1993). Developmental changes in short-term memory: A revised working memory perspective. In A. Collins (Ed.), *Theories of memory* (pp. 189–210). Hove, UK: Erlbaum.

Gathercole, S. E., Lamont, E., & Alloway, T. (2006). Working memory in the classroom. In S. J. Pickering (Ed.), *Working memory and education.* Elsevier Press.

Gathercole, S. E., & Pickering, S. J. (2000). Working memory deficits in children with low achievement in the national curriculum at 7 year of age. *British Journal of Educational Psychology, 70,* 177–194.

Gathercole, S. E., Pickering, S. J., Knight, C., & Stegmann, Z. (2004). Working memory skills and educational attainment: Evidence from national curriculum assessments at 7 and 14 years of age. *Applied Cognitive Psychology, 18,* 1–16.

Gathercole, S. E., Holmes, J., Alloway, T. P., Hilton, K. A., Place, M., & Elliott, J. E. (2010). Executive function deficits in children with low working memory. *In preparation.*

Geary, D. C. (1993). Mathematical disabilities: Cognitive, neuropsychological, and genetic components. *Psychological Bulletin, 114,* 345–362.

Geary, D. C. (2004). Mathematics and learning disabilities. *Journal of Learning Disabilities, 37,* 4–15.

Geary, D. C., Hoard, M. K., Byrd-Craven, J., & De Soto, M. C. (2004). Strategy choices in simple and complex addition: Contributions of working memory and counting knowledge for children with mathematical disability. *Journal of Experimental Child Psychology, 88,* 121–151.

Geary, D. C., Hoard, M. K., & Hamson, C. O. (1999). Numerical and arithmetical cognition: Patterns of functions and deficits in children at risk for a mathematical disability. *Journal of Experimental Child Psychology, 74,* 213–239.

Hamann, M. S., & Ashcraft, M. H. (1985). Simple and complex addition across development. *Journal of Experimental Child Psychology, 40,* 49–72.

Hoff, E., & Tian, C. (2005). Socioeconomic status and cultural influences on language. *Journal of Communication Disorders, 38,* 271–278.

Holmes, J., & Adams, J. W. (2006). Working memory and children's mathematical skills: Implications for mathematical development and mathematics curricula. *Educational Psychology, 26,* 339–366.

Holmes, J., Gathercole, S. E., & Dunning, D. L. (2009). Adaptive training leads to sustained enhancement of poor working memory in children. *Developmental Science, 12,* F9–F15.

Holmes, J., Gathercole, S. E., Place, M., Alloway, T. P., Elliott, J. G., & Hilton, K. A. (2010). The diagnostic utility of executive function assessments in the identification of ADHD in children. *Child & Adolescent Mental Health, 15,* 37–43.

Holmes, J., Gathercole, S. E., Place, M., Dunning, D. L., Hilton, K. A., & Elliott, J. G. (2010). Working memory deficits can be overcome: Impacts of training and medication on working memory in children with ADHD. *Applied Cognitive Psychology, 24,* 827–836.

Holmes, J., Dunning, D. L., & Gathercole, S. E. (2010a). Is working memory training for every child? *Manuscript in preparation.*

Holmes, J., Dunning, D. L., & Gathercole, S. E. (2010b). Impact of working memory training on language skills in children with dyslexia. *Manuscript in preparation.*

Holmes, J., Gathercole, S. E., Alloway, T. P., Hilton, K. A., Place, M., & Elliott, J.E. (2010). Working memory and executive function impairments in ADHD. *Manuscript under review.*

Hulme, C., & Muir, C. (1985). Developmental changes in speech rate and memory span-a causal relationship. *British Journal of Developmental Psychology, 3,* 175–181.

Huttenlocher, J., Haight, W., Bryk, A., Seltzer, M., & Lyons, T. (1991). Early vocabulary growth: Relation to language input and gender. *Developmental Psychology, 27,* 236–248.

Jaeggi, S. M., Buschkuel, M., Jonides, J., & Perrig, W. J. (2008). Improving fluid intelligence with training on working memory. *Proceedings of the National Academy of Sciences of the United States of America, 105,* 6829–6833.

Jarrold, C., Baddeley, A. D., & Hewes, A. K. (1999). Genetically dissociated components of working memory: Evidence from Down's and William's syndrome. *Neuropsychologia, 37,* 637–651.

Jarrold, C., Baddeley, A. D., Hewes, A. K., & Phillips, A. C. (2001). A longitudinal assessment of diverging verbal and non-verbal abilities in the Williams syndrome phenotype. *Cortex*, *37*, 423–431.

Jarvis, H. L., & Gathercole, S. E. (2003). Verbal and nonverbal working memory and achievements on national curriculum tests at 11 and 14 years of age. *Educational and Child Psychology*, *20*, 123–140.

Jeffries, S. A., & Everatt, J. (2003). Differences between dyspraxics and dyslexics in sequence learning and working memory. *Dyspraxia Foundation Professional Journal*, *2*, 12–21.

Jeffries, S. A., & Everatt, J. (2004). Working memory: Its role in dyslexia and other specific learning difficulties. *Dyslexia*, *10*, 196–214.

Jurden, F. H. (1995). Individual differences in working memory and complex cognition. *Journal of Educational Psychology*, *87*, 93–102.

Just, M. A., & Carpenter, P. A. (1992). A capacity theory of comprehension: Individual differences in working memory. *Psychological Review*, *99*, 122–149.

Kane, M. J., Brown, L. H., McVay, J. C., Silvia, P. J., Myin-Germeys, I., & Kwapil, T. R. (2007). For whom the mind wanders, and when: An experience sampling study of working memory and executive control in everyday life. *Psychological Science*, *18*, 614–621.

Kane, M. J., Conway, A. R. A., Hambrick, D. Z., & Engle, R. W. (2007). Variation in working memory capacity as variation in executive attention and control. In A. R. A. Conway, C. Jarrold, M. J. Kane & J. N. Towse (Eds.), *Variation in working memory* (pp. 21–48). New York: Oxford University Press.

Kane, M. J., & Engle, R. W. (2000). Working memory capacity, proactive interference, and divided attention: Limits on long-term memory retrieval. *Journal of Experimental Psychology: Learning, Memory and Cognition*, *26*, 333–358.

Kane, M. J., Hambrick, D. Z., & Conway, A. R. A. (2005). Working memory capacity and fluid intelligence are strongly related constructs: Comment on Ackerman, Beier, and Boyle (2005). *Psychological Bulletin*, *131*, 66–71.

Kane, M. J., Hambrick, D. Z., Tuholski, S. W., Wilheim, O., Payne, T. W., & Engle, R. W. (2004). The generality of working memory capacity: A latent variable approach to verbal and visuo-spatial memory span and reasoning. *Journal of Experimental Psychology: General*, *133*, 189–217.

Kaye, D. B. (1986). The development of mathematical cognition. *Cognitive Development*, *1*, 157–170.

Kintsch, W., Healy, A. F., Hegarty, M., Pennington, B. F., & Salthouse, T. A. (1999). Models of working memory: Eight questions and some general issues. In A. Miyake & P. Shah (Eds.), *Models of working memory* (pp. 412–441). Cambridge: Cambridge University Press.

Klingberg, T., Fernell, E., Olesen, P. J., Johnson, M., Gustafsson, P., Dahlstrom, K., et al. (2005). Computerised training of working memory in children with ADHD-a randomised, controlled trial. *Journal of the American Academy of Child and Adolescent Psychiatry*, *44*, 177–186.

Klingberg, T., Forssberg, H., & Westerberg, H. (2002). Increased brain activity in frontal and parietal cortex underlies the development of visuospatial working memory capacity during childhood. *Journal of Cognitive Neuroscience*, *14*, 1–10.

Kremen, W. S., Jacobsen, K. C., Xian, H., Eisen, S. A., Eaves, L. J., Tsuang, M. T., et al. (2007). Genetics of verbal working memory processes: A twin study of middle-aged men. *Neuropsychologia*, *21*, 569–580.

Kristofferson, M. W. (1972). Effects of practice on character-classification performance. *Canadian Journal of Psychology*, *26*, 54–60.

Kwon, H., Reiss, A. L., & Menon, V. (2002). Neural basis of protracted developmental changes in visuo-spatial working memory. *Proceedings of the National Academy of Sciences of the United States of America*, 99, 13336–13341.

Kyllonen, P. C., & Christal, R. E. (1990). Reasoning ability is (little more than) working memory capacity? *Intelligence*, 14, 389–433.

Lange, G., & Pierce, S. H. (1992). Memory strategy learning and maintenance in preschool children. *Developmental Psychology*, 28, 453–462.

Leedale, R., Singleton, C., & Thomas, K. (2004). *Memory booster (computer program and manual)*. Beverly, UK: Lucid Research.

Lidow, M. S., Williams, G. V., & Goldman-Rakic, P. S. (1998). The cerebral cortex: A case for a common site of action of antipsychotics. *Trends in Pharmacological Sciences*, 19, 136–140.

Lui, M., & Tannock, R. (2007). Working memory and inattention in a community sample of children. *Behavioral and Brain Functions*, 23, 3–12.

Marsh, H. W., & Craven, R. G. (2006). Reciprocal effects of self-concept and performance from a multidimensional perspective: Beyond seductive pleasure and unidimensional perspectives. *Perspectives on Psychological Science*, 1, 133–163.

Martinussen, R., Hayden, J., Hogg-Johnson, S., & Tannock, R. (2005). A meta-analysis of working memory impairments in children attention-deficit/hyperactivity disorder. *Journal of the American Academy of Child and Adolescent Psychiatry*, 44, 377–384.

Martinussen, R., & Tannock, R. (2006). Working memory impairments in children with attention-deficit hyperactivity disorder with and without comorbid language learning disorders. *Journal of Clinical and Experimental Neuropsychology*, 28, 1073–1094.

Mayringer, H., & Wimmer, H. (2000). Psedoname learning by German-speaking children with dyslexia: Evidence for a phonological learning deficit. *Journal of Experimental Child Psychology*, 75, 116–133.

McInnes, A., Humphries, T., Hogg-Johnson, S., & Tannock, R. (2003). Listening comprehension and working memory are impaired in attention-deficit/hyperactivity disorder irrespective of language impairment. *Journal of Abnormal Child Psychology*, 17, 37–53.

McLean, J. F., & Hitch, G. J. (1999). Working memory impairments in children with specific arithmetical learning difficulties. *Journal of Experimental Child Psychology*, 74, 240–260.

McNab, F., Leroux, G., Strand, F., Thorell, L. B., Bergman, S., & Klingberg, T. (2008). Common and unique components of inhibition and working memory: An fMRI, within subjects investigation. *Neuropsychologia*, 46, 2668–2682.

McNab, F., Varrone, A., Farde, L., Jucaite, A., Bystritsky, P., Forssberg, H., et al. (2009). Changes in cortical dopamine D1 receptor binding associated with cognitive training. *Science*, 323, 800–802.

McNamara, D. S., & Scott, J. L. (2001). Working memory capacity and strategy use. *Memory and Cognition*, 29, 10–17.

Mehta, M. A., Goodyear, I. M., & Sahakian, B. J. (2004). Methylphenidate improves working memory function and set-shifting AD/HD: Relationships to baseline memory capacity. *Journal of Child Psychology and Psychiatry*, 45, 293–305.

Miyake, A., Friedman, N. P., Emerson, M. J., Witzki, A. H., Howerter, A., & Wager, T. D. (2000). The unity and diversity of executive functions and their contributions to complex "frontal lobe" tasks: A latent variable analysis. *Cognitive Psychology*, 41, 49–100.

Miyake, A., Friedman, N. P., Shah, P., Rettinger, D. A., & Hegarty, M. (2001). How are visuo-spatial working memory, executive functioning, and spatial abilities related? A latent-variable analysis. *Journal of Experimental Psychology, General*, 130, 621–640.

Miyake, A., & Shah, P. (1999). *Models of working memory: Mechanisms of active maintenance and executive control*. New York: Cambridge University Press.

Monsell, S. (1996). Control of mental processes. In V. Bruce (Ed.), *Unsolved mysteries of the mind: Tutorial essays in cognition* (pp. 93–148). Hove, UK: Lawrence Erlbaum Associates Ltd.

Montgomery, J. (1995). Sentence comprehension in children with specific language impairment: The role of phonological working memory. *Journal of Speech and Hearing Research, 38*, 187–199.

Montgomery, J. (2000a). Relation of working memory to offline and real-time sentence processing in children with specific language impairment. *Applied Psycholinguistics, 21*, 117–148.

Montgomery, J. (2000b). Verbal working memory in sentence comprehension in children with specific language impairment. *Journal of Speech, Language, and Hearing Research, 43*, 293–308.

Moody, D. E. (2009). Can intelligence be increased by training on a task of working memory? *Intelligence, 37*, 327–328.

Morris, E. (2002). *Insight primary.* Hampshire, UK: NFER Nelson.

Numminen, H., Service, E., Ahonen, T., & Ruoppila, I. (2001). Working memory and everyday cognition in adults with Down's syndrome. *Journal of Intellectual Disability Research, 45*, 157–168.

Olesen, P., Westerberg, H., & Klingberg, T. (2004). Increased prefrontal and parietal brain activity after training of working memory. *Nature Neuroscience, 7*, 75–79.

Ornstein, P. A., Baker-Ward, L., & Naus, M. J. (1988). The development of mnemonic skill. In P. E. Weinert & M. Perlmutter (Eds.), *Memory development: Universal changes and individual differences* (pp. 31–50). Hillsdale, NJ: Erlbaum.

Ornstein, P. A., & Naus, M. J. (1985). Effects of the knowledge base on children's memory strategies. In H. W. Reese (Ed.), *Advances in child development and behavior* (pp. 113–152). San Diego, CA: Academic Press.

Ottem, E., Lian, A., & Karlsen, P. (2007). Reasons for the growth of traditional memory span across age. *European Journal of Cognitive Psychology, 19*, 233–270.

Palmer, S. (2000). Working memory: A developmental study of phonological recoding. *Memory, 8*, 179–193.

Passolunghi, M. C., & Siegel, L. S. (2004). Working memory and access to numerical information in children with disability in mathematics. *Journal of Experimental Child Psychology, 88*, 348–367.

Pennington, B. F., Moon, J., Edgin, J., Stedron, J., & Nadel, L. (2003). The neuropsychology of Down syndrome: Evidence for hippocampal dysfunction. *Child Development, 74*, 75–93.

Pennington, B. F., & Ozonoff, S. (1996). Executive functions and developmental psychopathology. *Journal of Child Psychology & Psychiatry & Allied Disciplines, 37*, 57–87.

Pessoa, L., Gutierrez, E., Bandettini, P. A., & Ungerleider, L. G. (2002). Neural correlates of visual working memory: fMRI amplitude predicts task performance. *Neuron, 35*, 975–987.

Phillips, C. J., & Nettlebeck, T. (1984). Effects of practice on recognition memory of mildly mentally retarded adults. *American Journal of Mental Deficiency, 88*, 678–687.

Pickering, S., & Gathercole, S. (2001). *Working memory test battery for children (WMTB-C).* London: The Psychological Corporation.

Pickering, S. J., & Gathercole, S. E. (2004). Distinctive working memory profiles in children with special educational needs. *Educational Psychology, 24*, 393–408.

Roberts, R. J., & Pennington, B. F. (1996). An interactive framework for examining prefrontal cognitive processes. *Developmental Neuropsychology, 12*, 105–126.

Rodriguez, M., & Sadoski, M. (2000). Effects of rote, context, keyword, and context/keyword methods on retention of vocabulary in EFL classrooms. *Language Learning, 50*, 385–412.

Roodenrys, S. (2006). Working memory function in attention deficit hyperactivity disorder. In T. P. Alloway & S. E. Gathercole (Eds.), *Working memory and neurodevelopmental disorders*. Hove and New York: Psychology Press.

Rypma, B., & D'Esposito, M. (2003). A subsequent-memory effect in dorsolateral prefrontal cortex. *Cognitive Brain Research, 16*, 162–166.

Santos, F. H., & Bueno, O. F. A. (2003). Validation of the Brazilian children's test of pseudoword repetition in Portuguese speakers aged 4 to 10 years. *Brazilian Journal of Medical and Biological Research, 36*, 1533–1547.

Siegel, L. S., & Ryan, E. B. (1989). The development of working memory in normally achieving and subtypes of learning disabled children. *Child Development, 60*, 973–980.

Snowling, M. J., Muter, V., & Carroll, J. M. (2007). Children at family risk of dyslexia: A follow-up in adolescence. *Journal of Child Psychology & Psychiatry, 48*, 609–618.

St. Clair-Thompson, H. L., & Gathercole, S. E. (2006). Executive functions and achievements on national curriculum tests: Shifting, updating, inhibition, and working memory. *Quarterly Journal of Experimental Psychology, 59*, 745–759.

St. Clair-Thompson, H. L., & Holmes, J. (2008). Improving short-term and working memory: Methods of memory training. In N. B. Johansen (Ed.), *New research on short-term memory* (pp. 125–154). New York: Nova Science.

St. Clair-Thompson, H. L., Stevens, R., Hunt, A., & Bolder, E. (2010). Improving children's working memory and classroom performance. *Educational Psychology, 30*, 203–220.

Stroop, J. R. (1935). Studies of interference in sérial verbal reactions. *Journal of Experimental Psychology, 18*, 643–662.

Swanson, H. L. (1993). Working memory in learning disability subgroups. *Journal of Experimental Child Psychology, 56*, 87–114.

Swanson, H. L. (2003). Age-related differences in learning disabled and skilled readers' working memory. *Journal of Experimental Child Psychology, 85*, 1–31.

Swanson, H. L., & Beebe-Frankenberger, M. (2004). The relationship between working memory and mathematical problem solving in children at risk and not at risk for math disabilities. *Journal of Educational Psychology, 96*, 471–491.

Swanson, H. L., & Jerman, O. (2007). The influence of working memory on reading growth in subgroups of children with reading disabilities. *Journal of Experimental Child Psychology, 96*, 249–283.

Swanson, H. L., & Sachse-Lee, C. (2001). Mathematical problem solving and working memory in children with learning disabilities: Both executive and phonological processes are important. *Journal of Experimental Child Psychology, 79*, 294–321.

Swanson, H. L., & Saez, L. (2003). Memory difficulties in children and adults with learning disabilities. In H. L. Swanson, S. Graham & K. R. Harris (Eds.), *Handbook of learning disabilities* (pp. 182–198). New York: Guildford Press.

Swanson, H. L., & Siegel, L. (2001). Learning disabilities as a working memory deficit. *Issues in Education: Contributions of Educational Psychology, 7*, 1–48.

Thorell, L. B., Lindqvist, S., Bergman, S., Bohlin, G., & Klingberg, T. (2009). Training and transfer effects of executive functions in preschool children. *Developmental Science, 12*, 106–113.

Turley-Ames, K. J., & Whitfield, M. M. (2003). Strategy training and working memory task performance. *Journal of Memory and Language, 49*, 446–468.

Vijayraghavan, S., Wang, M., Birnbaum, S. G., Williams, G. V., & Arnsten, A. F. (2007). Inverted-U dopamine D1 receptor actions on prefrontal neurons engaged in working memory. *Nature Neuroscience, 10*, 376–384.

Wechsler, D. (2004). *Wechsler intelligence scale for children* (4th ed.). Oxford: Pearson Assessment.

Westerberg, H., Jacobaeus, H., Hirvikoski, T., Clevberger, P., Ostensson, M. L., Bartfai, A., *et al.* (2007). Computerized working memory training after stroke—A pilot study. *Brain Injury, 21,* 21–29.

Westerberg, H., & Klingberg, T. (2007). Changes in cortical activity after training of working memory—A single-subject analysis. *Physiology & Behavior, 92,* 186–192.

Willcutt, E. G., Doyle, A. E., Nigg, J. T., Faraone, S. V., & Pennington, B. (2005). Validity of the executive function theory of attention-deficit/hyperactivity disorder: A meta-analytic review. *Biological Psychology, 57,* 1336–1346.

Willcutt, E. G., Pennington, B. F., Chhabildas, N. A., Olson, R. K., & Hulslander, J. L. (2005). Neuropsychological analyses of comorbidity between RD and ADHD: In search of the common deficit. *Developmental Neuropsychology, 27,* 35–78.

Williams, G. V., & Goldman-Rakic, P. S. (1995). Modulation of memory fields by dopamine D1 receptors in prefrontal cortex. *Nature, 376,* 572–575.

Wilson, J. T., Scott, J. H., & Power, K. G. (1987). Developmental differences in the span of visual memory for pattern. *British Journal of Developmental Psychology, 5,* 249–255.

Wilson, K. M., & Swanson, H. L. (2001). Are mathematical disabilities due to a domain-general or a domain-specific working memory deficit? *Journal of Learning Disabilities, 34,* 237–248.

MATHEMATICAL LEARNING DISABILITIES

David C. Geary

DEPARTMENT OF PSYCHOLOGICAL SCIENCES, UNIVERSITY OF MISSOURI,
COLUMBIA, MISSOURI, USA

I. Introduction

Basic competencies in arithmetic and simple algebra influence employ-ability, wages, and on-the-job productivity above and beyond the con-tributions of reading ability and intelligence (Rivera-Batiz, 1992), and entry into high-paying science and technology fields requires an even deeper understanding of mathematics (Paglin & Rufolo, 1990). These and many other studies confirm the individual and society-level benefits of a workforce with strong mathematical abilities (National Mathematics Advisory Panel, 2008), and in doing so highlight the long-term costs to people who have difficulties in learning mathematics. These are individuals who will be particularly disadvantaged in the workforce of the twenty-first century and in their ability to function in many now rou-tine day-to-day activities that require mathematical knowledge; as with

Advances in Child Development and Behavior
Patricia Bauer : Editor

literacy in the twentieth century, numeracy will be needed for everyday living in the twenty-first century (Every Child a Chance Trust, 2009). The early identification of children at risk for poor long-term outcomes in mathematics and interventions designed to address these risks is critical, because children who start kindergarten behind their peers tend to stay behind throughout their schooling (Duncan *et al.*, 2007).

During the past two decades, considerable progress has been made in identifying the pattern of cognitive deficits that underlie the slow progress of children with mathematical learning disability (MLD; Geary, Hoard, Byrd-Craven, Nugent, & Numtee, 2007; Jordan, Kaplan, Ramineni, & Locuniak, 2009; Murphy, Mazzocco, Hanich, & Early, 2007) and those of low achieving (LA) children, that is, children whose mathematical achievement is consistently below expectations based on their intelligence and reading achievement (Berch & Mazzocco, 2007). Progress has also been made on the development of screening measures and assessments for the identification of these children (Geary, Bailey, & Hoard, 2009; Locuniak & Jordan, 2008). Several intervention studies are in progress and should result in remediation strategies in the coming decade.

I begin with an overview of the basic characteristics of MLD and LA children, but my primary focus is on their cognitive profiles, including discussion of specific deficits in the domains of number, counting, and arithmetic and potential mediators (e.g., working memory) of these deficits. Although the literatures are much sparser in comparison to the cognitive studies, I briefly review the potential social and emotional characteristics of these children, followed by discussion of intervention research.

A. DEFINITION

The *Diagnostic and Statistical Manual of Mental Disorders* (American Psychiatric Association, 1994) defines MLD in terms of a discrepancy between performance on mathematics achievement tests and expected performance based on age, intelligence, and years of education. Although this is a commonsense approach to defining MLD, there is in fact no agreed upon test, achievement cutoff score, or achievement–intelligence discrepancy for defining or diagnosing MLD or LA (Gersten, Clarke, & Mazzocco, 2007; Mazzocco, 2007). A consensus is emerging, however, with respect to the importance of distinguishing between these two groups and the associated achievement patterns (Geary *et al.*, 2007; Murphy *et al.*, 2007). Children who score at or below the 10th percentile on standardized mathematics achievement tests for at least 2 consecutive academic years are categorized as MLD, at least in research studies, and children scoring between the 11th and the 25th percentiles, inclusive, across 2 consecutive years are categorized as LA.

B. INCIDENCE

On the basis of several population-based, long-term studies, and many smaller-scale studies, about 7% of children and adolescents will be diagnosable as MLD in at least one area of mathematics before graduating from high school (Barbaresi, Katusic, Colligan, Weaver, & Jacobsen, 2005; Lewis, Hitch, & Walker, 1994; Shalev, Manor, & Gross-Tsur, 2005), and an additional 5–10% of children and adolescents will be identified as LA (Berch & Mazzocco, 2007; Geary *et al.*, 2007; Murphy *et al.*, 2007).

These percentages are not in accord with above noted 10th percentile cutoff for MLD and 11–25th percentile for LA because these cutoffs are based on performance across more than 1 academic year. In any 1 year, 10% of children will score at or below the 10th percentile by definition, but not all of them will score in this range across multiple years; the same goes for scores in the LA range. The multiple year criteria is important in defining these groups because many children who score poorly in 1 academic year may score much higher in the next. Children with such fluctuating achievement levels do not have the cognitive deficits that are found with children who consistently score in the MLD and LA ranges (Geary, Brown, & Samaranayake, 1991; Geary, Hamson, & Hoard, 2000).

C. ETIOLOGY

As with other forms of learning disability, twin and family studies reveal genetic and environmental contributions to individual differences in mathematics achievement and to MLD and LA (Kovas, Haworth, Dale, & Plomin, 2007; Light & DeFries, 1995; Shalev *et al.*, 2001). Shalev and her colleagues found that family members of children with MLD are 10 times more likely to be diagnosed with MLD than are individuals in the general population. In a large study of elementary school twins, Kovas *et al.* found genetic as well as shared (between the pair of twins) and unique environmental contributions to individual differences in mathematics achievement and MLD, with the latter defined by cutoffs at the 5th and 15th percentiles. Depending on the grade and mathematics test used, from 1/2 to 2/3 of the individual variation in mathematics achievement was attributable to genetic variation and the remainder to shared and unique experiences.

The genetic influences responsible for the low performance associated with MLD, whether defined as below the 5th percentile or the 15th, were responsible for individual differences at all levels of performance (Kovas *et al.*, 2007; Oliver *et al.*, 2004). These results suggest there are not specific MLD genes, but rather the genetic influences on MLD are the same as those that influence mathematics achievement across the continuum of scores. Of the genetic effects, 1/3 were shared with intelligence, 1/3 with reading achievement independent of intelligence, and 1/3 were unique to mathematics. The implication is that roughly 2/3 of the genetic influences on mathematics achievement and MLD are the same as those that influence learning in other academic areas, and 1/3 of these genetic influences only effect mathematics learning. The genetic factors that influence achievement across academic domains may explain why many children with MLD have reading disability (RD) or other difficulties that interfere with learning in school, such as attention deficit hyperactivity disorder (ADHD; Barbaresi *et al.*, 2005; Fletcher, 2005; Shalev *et al.*, 2001). Barbaresi *et al.* found that between 57% and 64% of individuals with MLD also had RD, depending on the diagnostic criteria used for MLD.

I note that moderate genetic influences on MLD should not be equated with constraint on the potential to remediate these deficits, because changes in the individuals' environment may alter the relative extent of genetic and environmental influences on MLD status and/or related outcomes. In any event, the genetic studies also indicate important environmental influences on mathematics learning and MLD. For example, schooling influences mathematics achievement in general and interventions for MLD improve the mathematics achievement of these children above and beyond the influence of general education (Section IV), even if they do not eliminate individual differences.

II. Cognitive Profile

Cognitive studies of children with MLD and LA children have largely focused on number, counting, and arithmetic (Geary, 2004; Jordan & Montani, 1997; Russell & Ginsburg, 1984), and are supplemented by neuropsychological studies of developmental dyscalculia (Butterworth, 2005; Butterworth & Reigosa, 2007; Rourke, 1993; Temple, 1991). The latter includes children who have very specific deficits in number, counting, or arithmetic that are presumed to be due to a neurodevelopmental disorder. Most of these children would fall within the MLD category (Geary, 1993). The goals in both research traditions have been to identify the core deficits within these domains and to identify the mechanisms, such as working memory, that may mediate or moderate the expression of these deficits. The proposed mechanisms underlying the cognitive patterns described in the following sections range from a fundamental deficit in potentially inherent number and magnitude representational systems (Butterworth, 2005) to domain-general deficits in working memory that make the learning of mathematics, and other academic domains, in school difficult (e.g., Swanson, 1993; Swanson & Sachse-Lee, 2001).

A deficit in a fundamental magnitude representational system could affect the learning of mathematical content that is dependent on an understanding of numerical magnitude and thus cascade into many or all of the number, counting, and arithmetic deficits that have been identified in children with MLD and, to a lesser extent, LA children. The educational implications of a fundamental difficulty in understanding magnitude are very different from mathematical cognition deficits that result from below average working memory capacity, and thus studies aimed at identifying these mechanisms are critical to the advancement of the field (Gersten, Jordan, & Flojo, 2005).

A. NUMBER

1. Typical Development

Butterworth and colleagues have proposed that dyscalculia and presumably MLD result from deficits in two fundamental number sense systems, one that supports the representation and implicit understanding of the *exact quantity* of small collections of objects and of symbols (e.g., Arabic numerals) that represent these quantities (e.g., "3" = ■■■), and the other for representing the *approximate magnitude* of larger quantities (Butterworth, 2005; Butterworth & Reigosa, 2007). For typically achieving (TA) children, these competences are manifested in their ability to (a) subitize, that is, to apprehend the quantity of sets of 3–4 objects or actions without counting (Starkey & Cooper, 1980; Strauss & Curtis, 1984; Mandler & Shebo, 1982; Wynn, Bloom, & Chiang, 2002); (b) use nonverbal processes or counting to quantify small sets of objects and to add and subtract small quantities to and from these sets (Case & Okamoto, 1996; Levine, Jordan, & Huttenlocher, 1992; Starkey, 1992); and (c) estimate the relative magnitude of sets of objects beyond the subitizing range and estimate the results of simple numerical operations (Dehaene, 1997), for example, implicitly knowing that adding one item to a set of items results in "more" (Wynn, 1992).

Even infants are sensitive to the exact quantity of small sets of objects (e.g., ■■ vs. ■■■; Antell & Keating, 1983), and the maximum of this set size (e.g., 3 vs. 4) may increase for some children during the preschool years (Starkey, 1992). The approximate representational system is assessed by infants' ability to discriminate *more than* and *less than* comparing large collections of objects. Six month olds can discriminate sets that differ by a ratio of 2:1 (e.g., 16 objects>8 objects, but not 14>8), that is, when the larger quantity is 100% more than the smaller one. This ability improves rapidly, due to some combination of brain maturation and experience, during the preschool years such that 6 year olds can discriminate quantities that differ by 20% and reach the adult level of discrimination (12%) later in childhood (Halberda & Feigenson, 2008). These fundamental numerical competencies provide the foundation for many aspects of children's early mathematics learning (Geary, 2006, 2007). For instance, the exact representational system supports children's initial understanding that Arabic numerals and number words represent distinct quantities (e.g., ■■■ = "3" = "three") and the approximate system supports learning in other areas of basic mathematics (Gilmore, McCarthy, & Spelke, 2007; Geary, 2010).

Among these basic areas is the number line. Children's placements on a physical number line are hypothesized to be influenced by access to their approximate magnitude representational system. Placements that are influenced by use of this system result in a pattern that conforms to the natural logarithm of the number (Feigenson, Dehaene, & Spelke, 2004; Gallistel & Gelman, 1992); specifically, the placements are compressed for larger magnitudes such that the perceived distance, as an example, between 8 and 9 is smaller than the perceived distance between 2 and 3. With schooling, children who are TA quickly develop number line placements that conform to the linear mathematical system (Siegler & Booth, 2004), that is, the difference between two consecutive numbers is identical regardless of where they lie on the number line.

2. Children with MLD and LA

In keeping with Butterworth's (2005) hypothesis, children with MLD and, to a lesser extent, LA children may have deficits or developmental delays for both subitizing and the ability to represent approximate quantities (Geary, Hoard, Nugent, & Byrd-Craven, 2008; Halberda, Mazzocco, & Feigenson, 2008; Koontz & Berch, 1996; Landerl, Bevan, & Butterworth, 2003).

In the first of these studies, Koontz and Berch (1996) administered a variant of Posner, Boies, Eichelman, and Taylor's (1969) physical and name identity task to third and fourth graders with MLD and their TA peers. As an illustration, these children were asked to determine if combinations of Arabic numerals (e.g., 3–2), number sets (■■–■■), or numerals and sets were the same (2–■■) or different (3–■■). Confirming earlier findings (Mandler & Shebo, 1982), the TA children's reaction time patterns indicated fast and automatic subitizing for quantities of two and three, whether the code was an Arabic numeral or number set. The children with MLD showed fast access to representations for the quantity of two, but appeared to rely on counting to determine quantities of three. The results suggest that some children with MLD might not have an inherent representation for numerosities of three or the exact representational system does not reliably discriminate two and three.

Geary *et al.* (2007) developed the Number Sets Test to assess the fluency with which children process and add sets of objects and Arabic numerals to match a target number; for example, whether the combination "●● 3" matches the target of "5." The items are similar to those used by Koontz and Berch (1996), albeit some involve magnitudes up to 9. Fluency should be aided by rapid subitizing, rapid access to the magnitudes of small Arabic numerals, and the ability to add and compare and contrast these magnitudes. Receiver operating characteristic (ROC) analyses can be used to generate sensitivity (*d*-prime) and response bias measures. The former provides a measure of sensitivity to number combinations that match the target value, whereas the latter measures the child's tendency to circle items whether or not they match the target. The *d*-prime but not response bias score was found to be correlated with mathematics but not reading achievement, above and beyond the influence of intelligence, working memory, and mathematics and reading achievement scores from prior grades (Geary *et al.*, 2009).

Several studies have revealed that the *d*-prime scores of children with MLD children are consistent with Koontz and Berch's (1996) findings and, in comparison to their TA peers, suggest about a 3-year delay in the fluency of accessing representations of small exact quantities and in combining and recombining them (Geary *et al.*, 2009; Geary, Hoard, & Bailey, in press). Moreover, deficits on the Number Sets Test appear to be independent of the below described difficulties with arithmetic fact retrieval and execution of arithmetic procedures. The performance of LA children is between that of children with MLD and TA children but closer to that of TA children.

Geary *et al.* (2007, 2008) compared the number line performance of MLD, LA, and TA first and second graders. Group differences emerged using group medians, with trial-by-trail assessments of whether the placement was consistent with a log (suggesting reliance on the approximate magnitude system) or linear (suggesting learning of the mathematical number line) representation, and with several measures of absolute error.

The latter included the absolute difference between the child's placement and the correct placement independent of the representation guiding the placement, and absolute degree of error for linear and log trials (e.g., degree to which the placement differed from the predicted log placement for log trials). The results confirmed that TA children quickly learn the linear mathematical number line.

Children with MLD were more heavily dependent on the approximate representational system and consistent with Butterworth's (2005) hypothesis their representation of magnitude appeared to be more compressed than that of LA and TA children. The children with MLD did not appear to discriminate the magnitudes of smaller numerals as easily as the other children. This was the case even when these children used the approximate representational system which should make discriminations of smaller values (e.g., 1 vs. 4) easier than the use of the linear, mathematical representation. The performance of the LA children suggested a 1-year delay in the maturation of the approximate representational system, or in the construction of the linear representation from this system.

B. COUNTING KNOWLEDGE

1. Typical Development

Most children quickly learn to count by rote, and this in and of itself is not a useful indicator of MLD or LA status. Rather, the research in this area has focused on whether these children understand the core principles of counting. Gelman and Gallistel (1978) proposed that children's counting behavior is guided by five inherent and implicit principles that mature during the preschool years. Of these, the principles of one–one correspondence (only one number word is assigned to each counted object), stable order (order of number words must be invariant across counts), and cardinality (final number word represents the quantity of items) define the initial "how to count" rules that in turn provide the potentially inherent skeletal structure for children's emerging counting knowledge (Gelman & Meck, 1983). Whether or not there are inherent constraints to children's emergent counting knowledge, children make inductions about the basic characteristics of counting by observing others' counting behavior (Briars & Siegler, 1984; Fuson, 1988). One result is a belief that certain unessential features of counting are essential. For instance, counting has to proceed from left to right.

The assessment of children's counting knowledge is based on their sensitivity to violations of core principles or standard ways of counting. In one procedure, the child is asked to monitor a puppet's counting of objects and to tell the puppet if the count was "okay" or "not okay" (Briars & Siegler, 1984; Gelman & Meck, 1983). Sometimes the puppet counts correctly and others times the puppet violates one of Gelman and Gallistel's (1978) implicit principles or Briars and Siegler's unessential features. If the child detects a violation of a how to count principle, it is assumed that the child at least implicitly understands the principle. If the child states that correct counting from right to left, for instance, is okay, then the child knows that the standard left to right counting is unessential (i.e., you can count in other ways and still get the correct answer, as long as each item is tagged only once with a counting word). Children's knowledge of counting principles and unessential features of counting and sensitivity to violations of these principles and features (e.g., while watching a puppet count) emerge during the preschool years and mature during the early elementary school years but not in a straightforward way (LeFevre *et al.*, 2006).

Important nuances have emerged for children's sensitivity to unusual but correct ways of counting, such as counting the 1st, 3rd, and 5th item in a set of five and then counting the 2nd and 4th item. The result is correct but the method deviates from the standard left to right count of contiguous items. LeFevre *et al.* (2006) found that low ability first graders had higher scores on these types of counting tasks—they were less likely to say the answers were wrong—than did their average and high ability peers, a trend that was reversed in second grade. LeFevre and her colleagues argued that lower ability children tend to say all counts are correct, unless the error is very obvious, whereas children who are more sensitive to nuances in counting often identify unusual counts as incorrect. With experience in different ways of counting, these perceptive children quickly learn that irregular counts can be correct, if other rules (e.g., no item is double counted) are not violated.

2. Children with MLD and LA

Elementary school children with MLD and LA children understand most basic counting principles, but they are sometimes confused when counting deviates from the standard left to right counting of adjacent objects (Geary, Bow-Thomas, & Yao, 1992; Geary, Hoard, Byrd-Craven, & DeSoto, 2004). The mixed findings for these types of items may reflect LeFevre *et al.*'s (2006) findings of developmental change in sensitivity to nuances in counting. Either way, a more consistent finding is that children with MLD, but not LA children, fail to detect errors when the puppet double counts the first object in a counted set, that is, the object is counted "one, two." They detect these double counts and know they are wrong when the last item is counted twice, indicating they understand one–one correspondence. However, when the first item is double counted they have difficulty in retaining a notation of the counting error in working memory until the end of the count (Geary *et al.*, 2004; Hoard, Geary, & Hamson, 1999).

The forgetting of miscounts is potentially problematic for children's learning to use counting to solve arithmetic problems (below). Ohlsson and Rees (1991) predicted that children who are skilled at detecting counting errors would more readily learn to correct these miscounts and thus eventually commit fewer errors when using counting to solve arithmetic problems. The evidence for this prediction is mixed (Geary *et al.*, 1992, 2004), but detection of these double-counting errors may still be a good empirical indicator of risk for MLD (Geary *et al.*, 2007; Gersten *et al.*, 2005).

C. ARITHMETIC

1. Typical Development

By the time they enter kindergarten, most children have coordinated their number knowledge and counting skills with an implicit understanding of addition and subtraction, resulting in an ability to use number words to solve formal addition and subtraction problems (e.g., "How much is $3+2$?"; Groen & Resnick, 1977; Levine *et al.*, 1992; Siegler & Jenkins, 1989). Although children of this age will use a mix of problem-solving strategies, the most common approaches involve counting, sometimes using their fingers (finger counting strategy) and sometimes not using them (verbal counting strategy; Siegler & Shrager, 1984). The min and sum procedures are two common ways children count (Groen & Parkman, 1972), with or without their fingers. The min procedure involves stating the larger-valued addend and then counting a number of times equal to the value of the smaller addend. The sum procedure involves counting both addends starting from 1; the less common max procedure involves stating the smaller addend and counting the larger one.

The use of counting results in the development of memory representations of basic facts (Siegler & Shrager, 1984). Once formed, these long-term memory representations support the use of memory-based processes. The most common of these are *direct retrieval* of arithmetic facts and *decomposition*. With direct retrieval, children state an answer that is associated in long-term memory with the presented problem, such as stating "eight" when asked to solve $5+3$. Decomposition involves reconstructing the answer based on the retrieval of a partial sum; for example, $6+7$ might be solved by retrieving the answer to $6+6$ and then adding 1 to this partial sum. The general pattern of change is from use of the least sophisticated problem-solving procedures, such as sum counting, to the most efficient retrieval-based processes (Ashcraft, 1982). Development, however, is not simply a switch from use of less sophisticated to more sophisticated strategies. Rather, at any time children can use any of the many strategies they know to solve different problems; they may retrieve the answer to $3+1$ but count to solve $5+8$. What changes is the mix of strategies, with sophisticated ones used more often and less sophisticated ones less often (Siegler, 1996).

2. Children with MLD and LA

Application of the same methods developed to study TA children's strategy mix to the study of the arithmetic competencies of children with MLD and LA children has revealed many similarities and a few notable differences (e.g., Geary, 1990; Geary & Brown, 1991; Hanich, Jordan, Kaplan, & Dick, 2001; Jordan, Hanich, & Kaplan, 2003a; Jordan & Montani, 1997; Ostad, 1997). Children with MLD and LA children use the same types of problem-solving approaches as their TA peers, but differ in their procedural competence and in the ease of development of long-term memory representations of basic facts (Geary, 1993).

a. Procedural Competence. In comparison to their same-grade TA peers, children with MLD, and to a lesser extend LA children, commit more procedural errors when they solve simple arithmetic problems $(4+3)$, simple word problems, and complex arithmetic problems (e.g., 745–198; Geary *et al.*, 2007; Hanich *et al.*, 2001; Jordan, Hanich, & Kaplan, 2003b; Russell & Ginsburg, 1984). Even when these children do not commit errors, they tend to use developmentally immature procedures (Geary, 1990; Hanich *et al.*, 2001; Jordan *et al.*, 2003b; Raghubar *et al.*, 2009). As an example, during the solving of simple addition problems, first graders with MLD use their fingers to help them keep track of counting and use the sum procedure more frequently than their TA peers. LA children also use their fingers more often than TA children, but use the min procedure more frequently than do children with MLD (Geary *et al.*, 2004, 2007; Jordan *et al.*, 2003b; Ostad, 1998). For simple arithmetic, this translates into roughly a 2- to 3-year developmental delay for children with MLD and about a 1-year delay for LA children.

The deficits and delays in children with MLD and LA children when solving simple problems become more evident when they attempt to solve more complex ones (Fuchs & Fuchs, 2002, Jordan & Hanich, 2000). During the solving of multistep arithmetic problems, such as 45×12 or

126 + 537, Russell and Ginsburg (1984) found that fourth grade children with MLD committed more errors than their IQ-matched TA peers. The errors involved the misalignment of numbers while writing down partial answers or while carrying or borrowing from one column to the next. Raghubar *et al.* (2009) confirmed this finding and found that it was more pronounced for subtraction than for addition. Common subtraction errors included subtracting the larger number from the smaller one (e.g., 83 − 44 = 41), failing to decrement after borrowing from one column to the next (e.g., 92 − 14 = 88; the 90 was not decremented to 80), and borrowing across 0s (e.g., 900 − 111 = 899). These patterns were found for children with MLD and children with LA, regardless of their reading achievement.

b. Fact Retrieval. One of the best documented findings in this area is that children with MLD and a subset of LA children have persistent difficulties committing basic arithmetic facts to long-term memory or retrieving them once they are committed (Barrouillet, Fayol, & Lathuliére, 1997; Geary, 1990; Geary *et al.*, 2000, in press; Jordan *et al.*, 2003a). It is not that these children cannot memorize or retrieve any basic facts, but rather they evince persistent difference in the frequency with which they correctly retrieve them and in the pattern of retrieval errors.

Three different types of mechanisms have been proposed as the potential source of these retrieval difficulties. The first is a deficit in the ability to form phonetic based representations in long-term memory (Geary, 1993). Children's early reliance on counting when they are first learning to solve arithmetic problems implicates an early dependence on the phonetic and semantic representational systems of the language domain. Any disruption in the ability to represent or retrieve information from these systems should, in theory, result in difficulties in forming problem/answer associations for arithmetic problems during the act of counting as well as result in comorbid word retrieval problems during the act of reading. The work of Dehaene and his colleagues suggests that the retrieval of addition facts is indeed supported by a system of neural structures that appear to support phonetic and semantic representations and are engaged during incrementing processes, such as counting (Dehaene & Cohen, 1995, 1997). These findings need to be interpreted with caution, however, because they are largely based on studies of adults and the brain and cognitive systems that support early learning probably differ in important ways from those that support the same competence in adulthood (Ansari, 2010).

The second mechanism is a deficit in the ability to inhibit irrelevant associations from entering working memory during the process of retrieving facts from long-term memory (Barrouillet *et al.*, 1997). The latter form of retrieval deficit was first discovered by Barrouillet *et al.* (1997) and is based on the working memory model of Conway and Engle (1994). One way to assess this form of retrieval problem is to ask children to solve a series of arithmetic problems but instruct them to try and remember the answer and not use counting or any other procedure to get the answer (Jordan & Montani, 1997). If intrusions disrupt children's ability to retrieve the correct answer, then the corresponding retrieval errors should be associated with the numbers in the presented problem. These would include, for instance, the retrieval of 36 when trying to solve 6×5, or 8 when trying to solve $4 + 7$. The first is a table-related error, that is, the correct answer to a similar (6×6 in this case) problem in the multiplication table (Campbell, 1987; Campbell & Graham, 1985) and the second is a counting-string error, that is, the retrieved answer follows one of the addends in the counting string (8 follows 7; Siegler & Robinson, 1982).

Barrouillet *et al.* (1997) found frequent intrusions of table-related errors when adolescents with MLD solved simple multiplication problems, and Geary *et al.* (2000) found a similar pattern when a combined group of second graders with MLD/LA solved simple addition problems. The children in the latter group committed more retrieval errors than did their TA peers, and between 17% and 29% of these errors were counting-string associates of one of the addends. In a longitudinal study of MLD, Geary *et al.* (in press) administered the same task to MLD, LA, and TA children in second to fourth grade, inclusive. For a subgroup of LA children (hereafter LA-R; defined based on the high percentage of retrieval errors), 85% of their retrieved answers were incorrect in all three grades, with little across-grade improvement. For the remaining LA children, 55% of their retrieved answers were errors in second grade and this dropped to 37% by fourth. For the children with MLD, 78% of their retrieved answers were errors in second grade and this dropped to 59% by fourth. The TA children had the fewest errors; 37–34% across grades.

The most relevant finding was for the pattern of counting-string intrusions. These were rare among the TA children (5% of retrieval errors in second grade), more common among the LA children (9%) and were especially frequent among the LA-R (21%) and MLD (21%) children (Geary *et al.*, in press). Unlike most other tasks in which LA children outperform children with MLD, the LA-R children showed no across-grade drop in the percentage of intrusion errors, but the percentage dropped to 8% by fourth grade for the children with MLD.

The third mechanism is the noted earlier deficits or delays in the systems that support subitizing and approximate representations of magnitude (Butterworth, 2005). Children's early learning of arithmetic may be particularly dependent on the latter system, that is, their ability to estimate the approximate answer to simple arithmetic problems may be an important conceptual foundation for learning to solve formal arithmetic problems and eventually retrieve answers to them (Gilmore et al., 2007). In this view, retrieval deficits are secondary to a more basic deficit in the approximate representational system. Empirical evaluation will require longitudinal studies to determine if there is a relation between number processing deficits during the preschool years and retrieval deficits in the elementary school years. Geary et al.'s (in press) study does not address this issue, but they did examine the relation between the overall frequency of retrieval errors and intrusions and performance on the Number Sets Test and number line tasks described in Section IIA. The critical finding was that LA-R children and the remaining LA children had similar levels of deficit, in comparison to TA children, on the Number Sets Test that could not be attributed to the mechanisms described in Section IID, and at the same time these groups differed sharply in the overall frequency of retrieval intrusion errors ($d = 2.24$). In other words, deficits in the number processing abilities assessed by the Number Sets Test were not related to the frequency of intrusion errors, contra Butterworth's (2005) prediction.

Overall, it is clear that difficulties learning or retrieving basic arithmetic facts are a common and persistent difficulty for children with MLD and for a subset of LA children. Of the proposed mechanisms, the evidence is strongest for intrusion errors, that is, the retrieval deficits are related in part to the intrusion of related, but task irrelevant information into working memory when these children are attempting to remember arithmetic facts. Not all of their errors are due to intrusions, however, suggesting multiple mechanisms may be involved and that different children may have retrieval deficits for different reasons. Whether these alternative mechanisms involve the language system and number processing deficits remains to be determined.

D. DOMAIN-GENERAL DEFICITS

Learning disabilities are by definition determined by the child's performance on academic achievement tests and in school more generally, and thus a thorough analysis of the potential sources of these disabilities must include factors that have been shown to predict school achievement. The domain general learning abilities include general fluid intelligence, working memory, and processing speed. The relations among these are robustly debated (e.g., Ackerman, Beier, & Boyle, 2005; Carroll, 1993), with some scientists proposing that working memory and fluid intelligence, for instance, are largely one and the same (Colom, Rebollo, Palacios, Juan-Espinosa, & Kyllonen, 2004; Kyllonen & Christal, 1990), others arguing that processing speed is the core attribute underlying performance on working memory and intelligence measures (Jensen, 1998), and still others arguing that although performance on all of these measures is correlated, they each assess unique competencies that are potentially important for academic learning (Embretson, 1995; Gathercole, Alloway, Willis, & Adams, 2006; Jurden, 1995).

While we await resolution of these issues, we can use Carroll's (1993) hierarchical organization of these competences as a heuristic for reviewing the associated research. General intelligence is at the top and represents processes that affect learning across contexts and content (see Deary, 2000; Geary, 2005). Working memory and processing speed are at the second level, and are broad abilities that affect learning in many but not all domains. At the third level are more restricted domains of competence, including mathematics. Psychometric studies indicate that school mathematics includes at least two ability factors, Numerical Facility, which assesses competence in arithmetic, and Mathematical Reasoning which assess more abstract mathematical knowledge (Geary, 1994).

1. Intelligence

General intelligence is the best individual predictor of achievement across academic domains, including mathematics (e.g., Deary, Strand, Smith, & Fernandes, 2007; Jensen, 1998; Taub, Floyd, Keith, & McGrew, 2008; Walberg, 1984). One of the most comprehensive of these studies was a 5-year prospective investigation of more than 70,000 students from 11 to 16 years of age (Deary et al., 2007). Intelligence assessed at 11 years explained nearly 60% of the variation on national mathematics tests 5 years later. Intelligence is also highly heritable and there appears to be shared genes contributing to the correlation between intelligence and mathematics achievement (Kovas, Harlaar, Petrill, & Plomin, 2005). One possibility then is that the slow mathematical growth of children with MLD and their LA peers and the partial heritability of these disorders might be related to intelligence.

Although this may be a contributing factor for children with MLD, it cannot be the entire story nor is it likely to be the main theme of this story. As a group, children with MLD tend to have intelligence scores in the low average range (standard scores of 90–95), and the contribution of intelligence to their achievement pattern has not been fully established (Geary et al., 2007; Murphy et al., 2007). It is clear nevertheless that their

achievement in mathematics and many of their specific mathematical cognition deficits are far below their intellectual potential, and are related to
deficits in working memory and potentially in more specific cognitive systems that support mathematical learning above and beyond the contributions of intelligence. Intelligence is not a likely contributor to the
low mathematics achievement of LA children, because as a group these
children tend to be of average intelligence and reading ability. The combination suggests at least an average ability to learn in school.

2. Working Memory

a. Basics. Working memory represents the ability to hold a mental representation in mind while simultaneously engaging in other mental processes. The core component is the central executive, which is expressed
as attention-driven control of information represented in two core representational systems (Baddeley, 1986; Baddeley & Hitch, 1974; Cowan,
1995). These are a language-based phonological loop (Baddeley,
Gathercole, & Papagno, 1998) and a visuospatial sketch pad (Logie,
1995). Measures of intelligence and working memory, especially the central executive, are moderately to highly correlated (e.g., Ackerman *et al.*,
2005; Conway, Cowan, Bunting, Therriault, & Minkoff, 2002), but capture
independent components of ability. Performance on both types of
measures requires attentional and inhibitory control, but these
mechanisms appear to be more important for performance on tests of
the central executive than on tests of intelligence, especially if the latter
are not timed. Performance on measures of intelligence, in contrast, is
much more dependent on the ability to think logically and systematically
(Embretson, 1995).

The relation between working memory capacity and performance on
mathematics achievement tests and on specific mathematical cognition
tasks is well established (DeStefano & LeFevre, 2004; Geary *et al.*, 2007;
McLean & Hitch, 1999; Swanson & Sachse-Lee, 2001). Whether assessed
concurrently or one or more years earlier, the higher the capacity of the
central executive the better the performance on measures of mathematics
achievement and cognition (Bull, Espy, & Wiebe, 2008; Mazzocco &
Kover, 2007; Passolunghi, Vercelloni, & Schadee, 2007). The importance
of the phonological loop and visuospatial sketch pad varies with the complexity and content of the mathematics being assessed (Bull *et al.*, 2008;
Geary *et al.*, 2007). The phonological loop appears to be important for
processes that involve the articulation of numbers, as in counting
(Krajewski & Schneider, 2009), the solving of mathematical word problems (Swanson & Sachse-Lee, 2001), and may be related to arithmetic

fact retrieval (Fuchs, Fuchs, Compton, *et al.*, 2006; Geary, 1993). The visuospatial sketch pad appears to be involved in a broader number of mathematical domains, although many details remain to be filled in (De Smedt *et al.*, 2009; Geary, 2010; Swanson, Jerman, & Zheng, 2008).

b. Children with MLD and LA Children. Children with MLD are consistently found to have working memory deficits that contribute to their slow progress in learning mathematics above and beyond the contributions of intelligence and processing speed (Bull, Johnston, & Roy, 1999; Geary *et al.*, 2004, 2007; McLean & Hitch, 1999; Swanson, 1993; Swanson, Jerman, & Zheng, 2009; Swanson & Sachse-Lee, 2001). Although most of these children have deficits in all three working memory systems, their compromised central executive appears to be especially problematic (Bull *et al.*, 1999; Geary *et al.*, 2007; Swanson, 1993), but the interpretation of these findings is complicated by at least three subcomponents of the central executive, each of which may affect mathematical learning in different ways. These include competence at maintaining information in working memory, task switching, and inhibiting the retrieval of irrelevant information (Bull & Scerif, 2001; Murphy *et al.*, 2007; Passolunghi, Cornoldi, & De Liberto, 1999; Passolunghi & Siegel, 2004). To further complicate the issue, some studies have found that the working memory deficits of children with MLD are specific to dealing with numerical information (McLean & Hitch, 1999; Siegel & Ryan, 1989), but other studies have found more general deficits (Swanson, 1993; Geary *et al.*, 2007).

In any case, difficulties inhibiting the activation of irrelevant information in working memory have been independently related to poor mathematics achievement by Bull and Scerif (2001) and Passolunghi and colleagues (Passolunghi *et al.*, 1999; Passolunghi & Siegel, 2004). As noted in Section IIC, deficits in this component of the central executive may explain children with MLD's high frequency of intrusion errors during the act of arithmetic fact retrieval and may be a contributing factor to the comorbidity of MLD and RD in some children. Poor readers are less able to suppress context-irrelevant meanings of ambiguous words (e.g., river bank, bank teller), the meanings of similar sounding words (e.g., patients, patience), and retrieve more contextual information than is appropriate for the read passage (Gernsbacher, 1993).

The intrusion errors of the earlier described LA-R children are also consistent with such a deficit, but their central executive, intelligence, and reading scores were all average (Geary *et al.*, in press). However, the central executive measures used in this study primarily assessed the maintenance component of the central executive and not the inhibitory control component, and the reading measure may not have been sensitive to intrusion errors. McLean and Hitch (1999) compared a group LA children to age and ability-matched (based on raw mathematics achievement test scores) controls, and found evidence that the LA children had deficits in the task switching component of the central executive, but they did not distinguish between LA children with specific retrieval deficits and those without them.

In our studies, we have found that LA children have average scores on measures of the phonological loop and visuospatial sketch pad (Geary *et al.*, 2007, in press), but the group of LA children assessed by McLean and Hitch (1999) had deficits on one of the two visuospatial working memory tasks they administered. As noted, children with MLD have deficits in both of these working memory systems that in turn may contribute to their slow progress in specific areas of mathematics (Geary *et al.*, 2007, 2008). As an example, the poor visuospatial working memory of children with MLD partially mediated their poor performance, relative to IQ-match LA children, on the Number Sets Test and their larger error scores on the number line task. The children with MLD also committed more errors when using counting to solve simple addition problems and this effect was fully mediated by their deficit on the phonological loop.

There is clearly much that remains to be learned about the relation between the multiple components of working memory and individual differences in learning across different areas of mathematics in general and the contributions of these working memory systems to the poor achievement of children with MLD and their LA peers. At this point, we can conclude that children with MLD have pervasive deficits across all of the working memory systems that have been assessed, but our understanding of the relations between specific components of working memory and specific mathematical cognition deficits is in its infancy. Many LA children, in contrast, appear to have a normal phonological working memory, especially if reading achievement is average or better, and a normal ability to use the attentional control functions of the central executive to maintain information in working memory. Many of these children also appear to have an intact visuospatial working memory system, but a subset of them may have more subtle deficits. The most promising results suggest that LA children have subtle deficits in the inhibitory control and task switching components of the central executive (Geary *et al.*, 2007; Murphy *et al.*, 2007), but we await confirmation.

3. Processing Speed

a. Basics. Individual differences in processing speed are correlated with individual differences in intelligence and working memory (Jensen, 1998), and to individual differences in basic verbal (Hunt, Lunneborg, & Lewis,

1975) and quantitative abilities (Geary & Widaman, 1987, 1992). The faster the speed of processing the higher the scores on psychometric tests, although the strength of these relations is smaller than that found between intelligence, working memory, and achievement tests (Deary, 2000; Geary, 2005). The relation between working memory and processing speed is currently debated; specifically, whether individual differences in working memory are driven by more fundamental differences in speed of cognitive processing and decision making (Ackerman *et al.*, 2005), or whether the attentional focus associated with the central executive speeds information processing (Engle, Tuholski, Laughlin, & Conway, 1999).

Whatever the direction of the relation, processing speed itself has several subcomponents that appear to be independent of working memory (Carroll, 1993), and is sometimes found to be a better predictor of mathematics outcomes than working memory (Bull & Johnston, 1997) or an independent predictor after controlling for working memory and intelligence (Geary, 2010). Developmentally, processing speed increases rapidly for many simple tasks during the early elementary school years and then asymptotes to near adult levels in adolescence (Kail, 1991). The mechanisms underlying this pattern are not well understood but may reflect a combination of substantial improvements in attentional focus during this time as well as rapid increases in neuronal white matter (which speeds neural transmission) during this age range (Giedd *et al.*, 1999).

b. Children with MLD and LA Children. Children with MLD and LA children take more time to solve problems, on average, than do their TA peers (Garnett & Fleischner, 1983), but this in and of itself is not an indication of a slower fundamental processing speed (Geary, 1993). This is because the memory difficulties of many of these children result in the frequent use of slower procedures for problem-solving and averaging problem-solving time across strategies results in slower average times. Componential analyses, where regression models are fitted to reaction times to decompose processing into subcomponents, such as speed of encoding numbers and speed of implicitly counting have yielded mixed results (Geary, 1990, Geary & Brown, 1991; Geary *et al.*, 1991). Estimates of speed of counting during execution of the verbal counting strategy, for instance, sometimes suggest children with MLD are slower at implicit counting than their TA peers but sometimes yield no differences, although the former groups are consistently found to be slower at more basic processes, such as encoding numbers into working memory.

Use of rapid automatized naming (RAN; Denckla & Rudel, 1976) is a better approach to the question of whether children with MLD and LA have a slower fundamental speed of encoding and processing information. This is because these measures involve simple, over-learned processes, as such reciting letter names and number words, and thus are not confounded by different strategic approaches. Slower performance on RAN tasks is consistently related to lower reading achievement scores (Swanson, Trainin, Necoechea, & Hammill, 2003), potentially mediated by ease of encoding and representing language sounds in the phonological loop (e.g., Arnell, Joanisse, Klein, Busseri, & Tannock, 2009). Using these same measures, Geary *et al.* (2007) found that first graders with MLD were substantially slower at speed of letter and number naming than their TA peers ($d = 1.7$) and LA children were moderately slower ($d = 0.5$). Slow processing speed in turn was related to frequent errors when the children with MLD used counting to solve relatively complex (e.g., $17 + 6$) addition problems.

To control for the potential confound with the central executive, I conducted analyses of RAN letter and number naming speeds across first to fifth grade for the groups of TA, LA, LA-R (recall, these children have frequent intrusion errors during fact retrieval), and MLD children described in Geary *et al.* (in press). The corresponding age range covers the period in which speed of processing rapidly increases across cognitive tasks (Kail, 1991). I contrasted the three latter groups with the TA group, first without controlling for the central executive and then controlling for it. As first and second graders, the TA children were faster at letter and number naming then their peers in both LA groups, but this advantage disappeared by fifth grade. Controlling for the central executive resulted in nonsignificant processing speed differences comparing the TA and LA groups, but the slower processing of the LA-R children remained significant, especially in first grade. The children with MLD were slower at letter and number naming in first to fourth grade, whether or not the central executive was controlled. They were also slower in fifth grade, but the difference was small and not significant.

Overall, children with MLD and their LA peers have slower processing speeds in early childhood but close this gap by fifth grade. For some LA children, their slow processing may be related to the attentional control component of the central executive rather than to a more fundamental difference in processing speed *per se*. For children with MLD, in contrast, there may be a more fundamental difference in the mechanisms (e.g., white matter development) that support speed of information processing but any such difference appears to be more of a developmental delay than a persistent deficit.

III. Social and Behavioral Profile

In a comprehensive analysis of six large longitudinal data sets, Duncan *et al.* (2007) tracked the relation between family background, social–emotional factors, attentional control, intelligence, and academic skills at school entry (i.e., 5–6 years of age) and long-term achievement in mathematics and reading. The best predictor of mathematics achievement throughout schooling was entry-level mathematics skills. Early attentional skills also predicted later achievement, but the magnitude of this effect was less than 1/4 the magnitude of the effect for entry-level mathematics skills. Internalizing (e.g., anxiety) and externalizing (e.g., aggression) problems at school entry were not related to later achievement nor were more general measures of social skills. This analysis suggests that the early social and behavioral profile of children is not related to their long-term mathematics achievement.

At the same time, studies of children who have been identified with a specific learning disability typically reveal these children to have an array of social deficits (Swanson & Malone, 1992) and that children referred for evaluation due to severe emotional or behavioral problems in school are often classified as learning disabled (e.g., Lopez & Forness, 1996). Most of these studies, unfortunately, have focused on RD, but they may still be relevant to MLD. Swanson and Malone's meta-analysis indicated that, as a group and in comparison to TA children, children who are classified as learning disabled experience more social rejection, have poor social problem-solving skills, and are reported by others as being aggressive and immature, among other issues (*d*s from .21 to .98).

The discrepant results across Duncan *et al.*'s (2007) study and studies based on samples of children who have been identified as learning disabled in school settings indicate there is much work to be done in this area, and especially with respect to children with MLD. Tentatively, it would appear that social–emotional functioning does not significantly affect children's learning of mathematics in school in a causal manner, but that children with MLD may show a host of social and behavioral issues that are comorbid with their learning disability.

IV. Interventions

There are few scientifically validated treatment protocols designed specifically to address the mathematical cognition deficits of children with MLD and their LA peers. The National Mathematics Advisory Panel conducted a meta-analysis of high quality mathematics interventions for students with a learning disability broadly defined and found that direct, teacher-guided explicit instruction on how to solve a specific type of mathematics problem was the most effective intervention (Gersten *et al.*, 2008). The interventions were always for multiple sessions extending over several weeks to 6 months and resulted in large improvements in students' ability to solve mathematical word problems, computational arithmetic problems, and novel word and arithmetic problems (ds from .78 to 1.3).

Fuchs and her colleagues are developing cognitively-motivated interventions for children at risk for MLD and LA (e.g., Fuchs, Fuchs, Hamlett, *et al.*, 2006; Fuchs *et al.*, 2010). They are designing these interventions to specifically focus on the mathematical cognition deficits described earlier. As an example, Fuchs *et al.* (2010) developed an intervention to increase the frequency and accuracy with which children with MLD use the min counting procedure to solve addition problems and a corresponding procedure to solve subtraction problems. The intervention was one-on-one and involved forty-eight 20- to 30-min sessions that included explicit instruction on how to use min counting, illustrated with a number line, to solve addition and subtraction problems. For some of the children, the instruction was followed by deliberate practice; specifically, if the child could not answer a simple addition or subtraction problem correctly within 1 min, they were instructed to use min counting to solve the problem. Other children were provided the same instruction, but read numerals afterwards instead of engaging in deliberate practice. The combination of explicit instruction and deliberate practice of min counting resulted in improved competence in solving simple addition and subtraction problems and more complex problems in which simple ones were embedded.

Interventions that are focused specifically on improving working memory may be another promising route to academic remediation for children with MLD (Diamond, Barnett, Thomas & Munro, 2007; Holmes, Gathercole, Place, *et al.*, 2009; Klingberg *et al.*, 2005; Thorell, Lindqvist, Nutley, Bohlin, & Klingberg, 2009). These interventions typically involve asking children to engage in tasks that tax their working memory capacity, that is, tasks that require simultaneous processing and manipulation of information that is close to the maximum they can effectively handle. Holmes, Gathercole and Dunning (2009), for example, demonstrated that children who engaged in an intervention that matched task difficulty to their current working memory capacity, but not an easier intervention, showed large gains in verbal ($d = 1.55$) and visuospatial ($d = 1.03$) working memory following about 20 training sessions (35 min each), and retained these gains at a 6-month follow-up; $d = .85$ and .52 for verbal and visuospatial working memory, respectively. These children also showed a modest ($d = .49$) gain on a mathematical reasoning test at the 6-month follow-up. The source or sources of these gains are not fully understood, but may involve improved attentional control.

Several of these interventions have also focused on the inhibitory control component of working memory that may contribute to the fact-retrieval deficits of LA-R children, but the results for these interventions are mixed. Diamond *et al.* (2007) reported that their intervention resulted in improved inhibitory control for preschool children. Thorell *et al.* (2009) assessed several interventions specifically designed to improve inhibitory control and found improvements on the training tasks, but these gains did not generalize to nontraining inhibitory tasks. Despite the mixed results, these interventions and those that seem to improve attentional control hold considerable promise for addressing the deficits and delays of children with MLD, and in future research they might be fruitfully combined with interventions that target critical mathematical competencies (e.g., Fuchs *et al.*, 2010).

Although it remains to be determined if children with MLD and LA children have comorbid social and behavioral problems, interventions for social problems have been developed (e.g., coaching on how to start conversations). Kavale and Mostert's (2004) meta-analysis of the effectiveness of these interventions revealed significant but small (*d*s of about .20) improvements in a variety of social skills.

V. Summary and Future Directions

During the past two decades, there have been substantial advances in our understanding of the cognitive correlates and potential deficits underlying MLD and LA. Researchers in the field are moving toward a diagnostic cutoff for MLD at the 15th percentile on a mathematics achievement test for more than one grade, which effectively results in identifying children who score below the 10th percentile in most grades (e.g., Murphy *et al.*, 2007); it is critical that MLD and LA groups are identified using achievement scores across two or more grades (Geary, 1990; Geary *et al.*, 1991). The cutoff for LA children is typically below the 25th or 30th percentile across several grades, which effectively results in identifying children who score at about the 20th percentile in most grades (Geary *et al.*, 2007). Overall, about 7% and 10% of children will meet these MLD or LA criteria, respectively, at some point before completing high school (Barbaresi *et al.*, 2005).

The cognitive research on these children has focused on their basic competencies in understanding number and counting and on their arithmetic skills. Children with MLD and, to a lesser extent, LA children show a deficit or delay in their processing of numbers (e.g., accessing the quantity associated with "3"), learning of arithmetic procedures, and in

memorizing basic arithmetic facts (Berch & Mazzocco, 2007). Most of
these children understand basic counting concepts but are not as skilled
as TA children in their ability to monitor and retain counting errors in
working memory, which in turn might contribute to their use of developmentally immature procedures and frequent errors when using counting
to solve arithmetic problems (Geary *et al.*, 2004). These learning
difficulties are related in part to low average IQ (i.e., 90–95) and below
average working memory capacity for children with MLD but not for
LA children. The latter children largely appear to have a below average
facility in dealing with numbers (e.g., adding ●●● + 2 = ?), use immature
arithmetic procedures, and a subset of them have particular difficulty
retrieving basic facts from long-term memory.

   Whatever the underlying causes, the number processing and procedural
difficulties appear to be more of a developmental delay (improves across
grades) than a deficit (shows little grade-to-grade improvement), with
LA children lagging 1 year behind their TA peers and MLD children
about 3 years behind (e.g., Geary *et al.*, 2004; Geary *et al.*, 2008). The
difficulties remembering arithmetic facts are more persistent for children
with MLD and for a subset of children with LA (Geary *et al.*, in press;
Jordan *et al.*, 2003a). These deficits in turn might be related to a poor ability to inhibit irrelevant information from intruding into working memory
during the act of retrieval, although this is not likely to be the only source
of their fact-retrieval deficits.

   Further advances in our understanding of the cognitive mechanisms
underlying the developmental delays and deficits of MLD and LA children will require not only fine grain assessment of the targeted mathematical competence but also fine grain assessment of potential mediators of
these deficits. Of particular interest will be studies that simultaneously
assess the multiple subsystems of the central executive component of
working memory and that relate these to the mathematical deficits and
delays of groups of MLD and LA children. Our understanding of these
disorders will also require expansion of the mathematical content domains
under study, with a particular emphasis on fractions, at least in the
elementary school years; algebra in high school. The National Mathematics Advisory Panel (2008) identified conceptual and procedural competencies in fractions as critical foundational skills for later mathematics
learning. We know that children in the United States have poorly developed fractions skills, and it is likely that children with MLD and LA will
be particularly disadvantaged (Mazzocco & Devlin, 2008). On the basis
of Duncan *et al.*'s (2007) results, early identification of children at risk
for later MLD and LA is critical, as children who start school behind their
peers in mathematics tend to stay behind. Some advances have been made

in this regard, but these studies have focused on mathematical cognition in kindergarteners and first graders (Geary *et al.*, 2009; Locuniak & Jordan, 2008). These studies need to be extended to the preschool years.

In recent years, we have seen several well-designed and promising intervention studies that focus on the specific mathematical cognition delays and deficits of children with MLD (e.g., Fuchs *et al.*, 2010), as well as more general interventions for improving working memory capacity (e.g., Holmes *et al.* 2009), but these are only the beginning of what is likely to be a very long journey in the development of cognitively informed treatment protocols. One area in which there has been little or if any progress is with regard to the social and emotional functioning of children with MLD and their LA peers. Studies of children with RD suggest heighted risk for comorbid social and emotional problems, but otherwise we know little about these issues. A final task for coming decades is to more fully explore the sources of the comorbidity of MLD, RD, and other disorders that affect learning. We know that comorbid disorders are common in these children, but we do not understand why this is the case (Fletcher, 2005).

Acknowledgments

During preparation of this chapter, the authors were supported by grant R37 HD045914 from the National Institute of Child Health and Human Development (NICHD).

REFERENCES

Ackerman, P. L., Beier, M. E., & Boyle, M. O. (2005). Working memory and intelligence: The same or different constructs? *Psychological Bulletin, 131*, 30–60.

American Psychiatric Association, (1994). *Diagnostic and statistical manual of mental disorders* (4th ed.). Washington, DC: American Psychiatric Association.

Ansari, D. (2010). Neurocognitive approaches to developmental disorders of numerical and mathematical cognition: The perils of neglecting development. *Learning and Individual Differences, 20*, 123–129.

Antell, S. E., & Keating, D. P. (1983). Perception of numerical invariance in neonates. *Child Development, 54*, 695–701.

Arnell, K. M., Joanisse, M. F., Klein, R. M., Busseri, M. A., & Tannock, R. (2009). Decomposing the relation between rapid automatized naming (RAN) and reading ability. *Canadian Journal of Experimental Psychology, 63*, 173–184.

Ashcraft, M. H. (1982). The development of mental arithmetic: A chronometric approach. *Developmental Review, 2*, 213–236.

Baddeley, A. D. (1986). *Working memory.* Oxford: Oxford University Press.

Baddeley, A., Gathercole, S., & Papagno, C. (1998). The phonological loop as a language learning device. *Psychological Review, 105,* 158–173.

Baddeley, A. D., & Hitch, G. J. (1974). Working memory. In G. H. Bower (Ed.), *The psychology of learning and motivation: Advances in research and theory* (Vol. 8, pp. 47–90). New York: Academic Press.

Barbaresi, W. J., Katusic, S. K., Colligan, R. C., Weaver, A. L., & Jacobsen, S. J. (2005). Math learning disorder: Incidence in a population-based birth cohort, 1976–82, Rochester, Minn. *Ambulatory Pediatrics, 5,* 281–289.

Barrouillet, P., Fayol, M., & Lathuliére, E. (1997). Selecting between competitors in multiplication tasks: An explanation of the errors produced by adolescents with learning disabilities. *International Journal of Behavioral Development, 21,* 253–275.

Berch, D. B., & Mazzocco, M. M. M. (Eds.), (2007). *Why is Math so Hard for Some Children? The Nature and Origins of Mathematical Learning Difficulties and Disabilities.* Baltimore, MD: Paul H. Brookes Publishing Co.

Briars, D., & Siegler, R. S. (1984). A featural analysis of preschoolers' counting knowledge. *Developmental Psychology, 20,* 607–618.

Bull, R., Espy, K. A., & Wiebe, S. A. (2008). Short-term memory, working memory, and executive functions in preschoolers: Longitudinal predictors of mathematical achievement at age 7 years. *Developmental Neuropsychology, 33,* 205–228.

Bull, R., & Johnston, R. S. (1997). Children's arithmetical difficulties: Contributions from processing speed, item identification, and short-term memory. *Journal of Experimental Child Psychology, 65,* 1–24.

Bull, R., Johnston, R. S., & Roy, J. A. (1999). Exploring the roles of the visual-spatial sketch pad and central executive in children's arithmetical skills: Views from cognition and developmental neuropsychology. *Developmental Neuropsychology, 15,* 421–442.

Bull, R., & Scerif, G. (2001). Executive functioning as a predictor of children's mathematical abilities: Inhibition, switching, and working memory. *Developmental Neuropsychology, 19,* 273–293.

Butterworth, B. (2005). Developmental dyscalculia. In J. I. D. Campbell (Ed.), *Handbook of Mathematical Cognition* (pp. 455–467). New York: Psychology Press.

Butterworth, B., & Reigosa, V. (2007). Information processing deficits in dyscalculia. In D. B. Berch & M. M. M. Mazzocco (Eds.), *Why is Math so Hard for Some Children? The Nature and Origins of Mathematical Learning Difficulties and Disabilities* (pp. 65–81). Baltimore, MD: Paul H. Brookes Publishing Co..

Campbell, J. I. D. (1987). Network interference and mental multiplication. *Journal of Experimental Psychology Learning, Memory, and Cognition, 13,* 109–123.

Campbell, J. I. D., & Graham, D. J. (1985). Mental multiplication skill: Structure, process, and acquisition. *Canadian Journal of Psychology, 39,* 338–366.

Carroll, J. B. (1993). *Human Cognitive Abilities: A survey of factor-analytic studies.* New York: Cambridge University Press.

Case, R., & Okamoto, Y. (1996). The role of central conceptual structures in the development of children's thought. *Monographs of the Society for Research in Child Development, 66,* (1–2, Serial No. 246).

Colom, R., Rebollo, I., Palacios, A., Juan-Espinosa, M., & Kyllonen, P. C. (2004). Working memory is (almost) perfectly predicted by *g. Intelligence, 32,* 277–296.

Conway, A. R. A., Cowan, N., Bunting, M. F., Therriault, D. J., & Minkoff, S. R. B. (2002). A latent variable analysis of working memory capacity, short-term memory capacity, processing speed, and general fluid intelligence. *Intelligence, 30,* 163–183.

Conway, A. R. A., & Engle, R. W. (1994). Working memory and retrieval: A resource-dependent inhibition model. *Journal of Experimental Psychology: General, 123,* 354–373.

Cowan, N. (1995). *Attention and memory: An integrated framework.* New York: Oxford University Press.

De Smedt, B., Janssen, R., Bouwens, K., Verschaffel, L., Boets, B., & Ghesquière, P. (2009). Working memory and individual differences in mathematics achievement: A longitudinal study from first grade to second grade. *Journal of Experimental Child Psychology, 103,* 186–201.

Deary, I. J. (2000). *Looking down on human intelligence: From psychophysics to the brain.* Oxford, UK: Oxford University Press.

Deary, I. J., Strand, S., Smith, P., & Fernandes, C. (2007). Intelligence and educational achievement. *Intelligence, 35,* 13–21.

Dehaene, S. (1997). *The number sense: How the mind creates mathematics.* NY: Oxford University Press.

Dehaene, S., & Cohen, L. (1995). Towards an anatomical and functional model of number processing. *Mathematical Cognition, 1,* 83–120.

Dehaene, S., & Cohen, L. (1997). Cerebral pathways for calculation: Double dissociation between rote verbal and quantitative knowledge of arithmetic. *Cortex, 33,* 219–250.

Denckla, M. B., & Rudel, R. (1976). Rapid automatized naming (RAN): Dyslexia differentiated from other learning disabilities. *Neuropsychologia, 14,* 471–479.

DeStefano, D., & LeFevre, J.-A. (2004). The role of working memory in mental arithmetic. *European Journal of Cognitive Psychology, 16,* 353–386.

Diamond, A., Barnett, S., Thomas, J., & Munro, S. (2007). Preschool program improves cognitive control. *Science, 318,* 1387–1388.

Duncan, G. J., Dowsett, C. J., Claessens, A., Magnuson, K., Huston, A. C., Klebanov, P., et al. (2007). School readiness and later achievement. *Developmental Psychology, 43,* 1428–1446.

Embretson, S. E. (1995). The role of working memory capacity and general control processes in intelligence. *Intelligence, 20,* 169–189.

Engle, R. W., Tuholski, S. W., Laughlin, J. E., & Conway, A. R. A. (1999). Working memory, short-term memory, and general fluid intelligence: A latent-variable approach. *Journal of Experimental Psychology: General, 128,* 309–331.

Every Child a Chance Trust, (2009). *The long-term costs of numeracy difficulties.* http://www.everychildachancetrust.org/counts/index.cfm. Retrieved August 14, 2009.

Feigenson, L., Dehaene, S., & Spelke, E. (2004). Core systems of number. *Trends in Cognitive Sciences, 8,* 307–314.

Fletcher, J. M. (2005). Predicting math outcomes: Reading predictors and comorbidity. *Journal of Learning Disabilities, 38,* 308–312.

Fuchs, L. S., & Fuchs, D. (2002). Mathematical problem-solving profiles of students with mathematics disabilities with and without comorbid reading disabilities. *Journal of Learning Disabilities, 35,* 563–573.

Fuchs, L. S., Fuchs, D., Compton, D. L., Powell, S. R., Seethaler, P. M., Capizzi, A. M., et al. (2006). The cognitive correlates of third-grade skill in arithmetic, algorithmic computation, and arithmetic word problems. *Journal of Educational Psychology, 98,* 29–43.

Fuchs, L. S., Fuchs, D., Hamlett, C. L., Powell, S. R., Capizzi, A. M., & Seethaler, P. M. (2006). The effects of computer-assisted instruction on number combination skill in at-risk first graders. *Journal of Learning Disabilities, 39,* 467–475.

Fuchs, L. S., Powell, S. R., Seethaler, P. M., Cirino, P. T., Fletcher, J. M., Fuchs, D., et al. (2010). The effects of strategic counting instruction, with and without deliberate practice, on number combination skill among students with mathematics difficulties. *Learning and Individual Differences, 20,* 89–100.

Fuson, K. C. (1988). *Children's counting and concepts of number.* New York: Springer-Verlag.

Gallistel, C. R., & Gelman, R. (1992). Preverbal and verbal counting and computation. *Cognition, 44,* 43–74.

Garnett, K., & Fleischner, J. E. (1983). Automatization and basic fact performance of normal and learning disabled children. *Learning Disabilities Quarterly, 6,* 223–230.

Gathercole, S. E., Alloway, T. P., Willis, C., & Adams, A.-M. (2006). Working memory in children with reading disabilities. *Journal of Experimental Child Psychology, 93,* 265–281.

Geary, D. C. (1990). A componential analysis of an early learning deficit in mathematics. *Journal of Experimental Child Psychology, 49,* 363–383.

Geary, D. C. (1993). Mathematical disabilities: Cognitive, neuropsychological, and genetic components. *Psychological Bulletin, 114,* 345–362.

Geary, D. C. (1994). *Children's mathematical development: Research and practical applications.* Washington, DC: American Psychological Association.

Geary, D. C. (2004). Mathematics and learning disabilities. *Journal of Learning Disabilities, 37,* 4–15.

Geary, D. C. (2005). *The origin of mind: Evolution of brain, cognition, and general intelligence.* Washington, DC: American Psychological Association.

Geary, D. C. (2006). Development of mathematical understanding. In D. Kuhl & R. S. Siegler (Eds.), *Cognition, Perception, and Language* (Vol. 2, pp. 777–810). New York: John Wiley & Sons W. Damon (Gen. Ed.), *Handbook of Child Psychology* (6th ed.).

Geary, D. C. (2007). An evolutionary perspective on learning disability in mathematics. *Developmental Neuropsychology, 32,* 471–519.

Geary, D. C. (2010). *Cognitive predictors of individual differences in achievement growth in mathematics and reading: A five year longitudinal study.* Under Editorial Review.

Geary, D. C., Hoard, M. K., & Bailey, D. H. (in press). Fact retrieval deficits in low achieving children and children with mathematical learning disability. *Journal of Learning Disability.*

Geary, D. C., Bailey, D. H., & Hoard, M. K. (2009). Predicting mathematical achievement and mathematical learning disability with a simple screening tool: The number sets test. *Journal of Psychoeducational Assessment, 27,* 265–279.

Geary, D. C., Bow-Thomas, C. C., & Yao, Y. (1992). Counting knowledge and skill in cognitive addition: A comparison of normal and mathematically disabled children. *Journal of Experimental Child Psychology, 54,* 372–391.

Geary, D. C., & Brown, S. C. (1991). Cognitive addition: Strategy choice and speed-of-processing differences in gifted, normal, and mathematically disabled children. *Developmental Psychology, 27,* 398–406.

Geary, D. C., Brown, S. C., & Samaranayake, V. A. (1991). Cognitive addition: A short longitudinal study of strategy choice and speed-of-processing differences in normal and mathematically disabled children. *Developmental Psychology, 27,* 787–797.

Geary, D. C., Hamson, C. O., & Hoard, M. K. (2000). Numerical and arithmetical cognition: A longitudinal study of process and concept deficits in children with learning disability. *Journal of Experimental Child Psychology, 77,* 236–263.

Geary, D. C., Hoard, M. K., Byrd-Craven, J., & DeSoto, C. M. (2004). Strategy choices in simple and complex addition: Contributions of working memory and counting knowledge for children with mathematical disability. *Journal of Experimental Child Psychology, 74,* 213–239.

Geary, D. C., Hoard, M. K., Byrd-Craven, J., Nugent, L., & Numtee, C. (2007). Cognitive mechanisms underlying achievement deficits in children with mathematical learning disability. *Child Development, 78,* 1343–1359.

Geary, D. C., Hoard, M. K., Nugent, L., & Byrd-Craven, J. (2008). Development of number line representations in children with mathematical learning disability. *Developmental Neuropsychology, 33,* 277–299.

Geary, D. C., & Widaman, K. F. (1987). Individual differences in cognitive arithmetic. *Journal of Experimental Psychology: General, 116,* 154–171.

Geary, D. C., & Widaman, K. F. (1992). Numerical cognition: On the convergence of componential and psychometric models. *Intelligence, 16,* 47–80.

Gelman, R., & Gallistel, C. R. (1978). *The child's understanding of number.* Cambridge, Mass: Harvard University Press.

Gelman, R., & Meck, E. (1983). Preschooler's counting: Principles before skill. *Cognition, 13,* 343–359.

Gernsbacher, M. A. (1993). Less skilled readers have less efficient suppression mechanisms. *Psychological Science, 4,* 294–298.

Gersten, R., Clarke, B., & Mazzocco, M. M. M. (2007). Historical and contemporary perspectives on mathematical learning disabilities. In D. B. Berch & M. M. M. Mazzocco (Eds.), *Why is Math so Hard for Some Children? The Nature and Origins of Mathematical Learning Difficulties and Disabilities* (pp. 7–28). Baltimore, MD: Paul H. Brookes Publishing Co..

Gersten, R., Ferrini-Mundy, J., Benbow, C., Clements, D. H., Loveless, T., Williams, V., et al. (2008). Report of the task group on instructional practices. In *National Mathematics Advisory Panel, Reports of the Task Groups and Subcommittees* (pp. 6-i–6-224). Washington, DC: United States Department of Education.

Gersten, R., Jordan, N. C., & Flojo, J. R. (2005). Early identification and interventions for students with mathematics difficulties. *Journal of Learning Disabilities, 38,* 293–304.

Giedd, J. N., Blumenthal, J., Jeffries, N. O., Castellanos, F. X., Liu, H., Zijdenbos, A., et al. (1999). Brain development during childhood and adolescence: A longitudinal MRI study. *Nature Neuroscience, 2,* 861–863.

Gilmore, C. K., McCarthy, S. E., & Spelke, E. S. (2007). Symbolic arithmetic knowledge without instruction. *Nature, 447,* 589–591.

Groen, G. J., & Parkman, J. M. (1972). A chronometric analysis of simple addition. *Psychological Review, 79,* 329–343.

Groen, G., & Resnick, L. B. (1977). Can preschool children invent addition algorithms? *Journal of Educational Psychology, 69,* 645–652.

Halberda, J., & Feigenson, L. (2008). Developmental change in the acuity of the "number sense": The approximate number system in 3-, 4-, 5-, and 6-year-olds and adults. *Developmental Psychology, 44,* 1457–1465.

Halberda, J., Mazzocco, M. M. M., & Feigenson, L. (2008). Individual differences in non-verbal number acuity correlate with maths achievement. *Nature, 455,* 665–668.

Hanich, L. B., Jordan, N. C., Kaplan, D., & Dick, J. (2001). Performance across different areas of mathematical cognition in children with learning difficulties. *Journal of Educational Psychology, 93,* 615–626.

Hoard, M. K., Geary, D. C., & Hamson, C. O. (1999). Numerical and arithmetical cognition: Performance of low- and average-IQ children. *Mathematical Cognition, 5,* 65–91.

Holmes, J., Gathercole, S. E., & Dunning, D. L. (2009). Adaptive training leads to sustained enhancement of poor working memory in children. *Developmental Science, 12,* F9–F15.

Holmes, J., Gathercole, S. E., Place, M., Dunning, D. L., Hilton, K. A., & Elliott, J. G. (2009). Working memory deficits can be overcome: Impacts of training and medication on working memory in children with ADHD. *Applied Cognitive Psychology,* doi:10.1002/acp. 1589.

Hunt, E., Lunneborg, C., & Lewis, J. (1975). What does it mean to be high verbal? *Cognitive Psychology, 7,* 194–227.

Jensen, A. R. (1998). *The g factor: The science of mental ability.* Westport, CT: Praeger.

Jordan, N., & Hanich, L. (2000). Mathematical thinking in second grade children with different forms of LD. *Journal of Learning Disabilities, 33,* 567–578.

Jordan, N. C., Hanich, L. B., & Kaplan, D. (2003a). Arithmetic fact mastery in young children: A longitudinal investigation. *Journal of Experimental Child Psychology, 85,* 103–119.

Jordan, N. C., Hanich, L. B., & Kaplan, D. (2003b). A longitudinal study of mathematical competencies in children with specific mathematics difficulties versus children with comorbid mathematics and reading difficulties. *Child Development, 74,* 834–850.

Jordan, N. C., Kaplan, D., Ramineni, C., & Locuniak, M. N. (2009). Early math matters: Kindergarten number competence and later mathematics outcomes. *Developmental Psychology, 45,* 850–867.

Jordan, N. C., & Montani, T. O. (1997). Cognitive arithmetic and problem solving: A comparison of children with specific and general mathematics difficulties. *Journal of Learning Disabilities, 30,* 624–634.

Jurden, F. H. (1995). Individual differences in working memory and complex cognition. *Journal of Educational Psychology, 87,* 93–102.

Kail, R. (1991). Developmental change in speed of processing during childhood and adolescence. *Psychological Bulletin, 109,* 490–501.

Kavale, K. A., & Mostert, M. P. (2004). Social skills interventions for individuals with learning disabilities. *Learning Disability Quarterly, 27,* 31–43.

Klingberg, T., Fernell, E., Olesen, P. J., Johnson, M., Gustafsson, P., Dahlström, K., et al. (2005). Computerized training of working memory in children with ADHD—A randomised, controlled trial. *Journal of the American Academy of Child and Adolescent Psychiatry, 44,* 177–186.

Koontz, K. L., & Berch, D. B. (1996). Identifying simple numerical stimuli: Processing inefficiencies exhibited by arithmetic learning disabled children. *Mathematical Cognition, 2,* 1–23.

Kovas, Y., Harlaar, H., Petrill, S. A., & Plomin, R. (2005). 'Generalist genes' and mathematics in 7-year-old twins. *Intelligence, 33,* 473–489.

Kovas, Y., Haworth, C. M. A., Dale, P. S., & Plomin, R. (2007). The genetic and environmental origins of learning abilities and disabilities in the early school years. *Monographs of the Society for Research in Child Development, 72,* (3, Serial No. 288).

Krajewski, K., & Schneider, W. (2009). Exploring the impact of phonological awareness, visual–spatial working memory, and preschool quantity–number competencies on mathematics achievement in elementary school: Findings from a 3-year longitudinal study. *Journal of Experimental Child Psychology, 103,* 516–531.

Kyllonen, P. C., & Christal, R. E. (1990). Reasoning ability is (Little more than) working-memory capacity?!. *Intelligence, 14,* 389–433.

Landerl, K., Bevan, A., & Butterworth, B. (2003). Developmental dyscalculia and basic numerical capacities: A study of 8–9-year-old students. *Cognition, 93,* 99–125.

LeFevre, J.-A., Smith-Chant, B. L., Fast, L., Skwarchuk, S.-L., Sargla, E., Arnup, J. S., et al. (2006). What counts as knowing? The development of conceptual and procedural knowledge of counting from kindergarten through Grade 2. *Journal of Experimental Child Psychology, 93,* 285–303.

Levine, S. C., Jordan, N. C., & Huttenlocher, J. (1992). Development of calculation abilities in young children. *Journal of Experimental Child Psychology, 53,* 72–103.

Lewis, C., Hitch, G. J., & Walker, P. (1994). The prevalence of specific arithmetic difficulties and specific reading difficulties in 9-year-old to 10-year-old boys and girls. *Journal of Child Psychology and Psychiatry, 35*, 283–292.

Light, J. G., & DeFries, J. C. (1995). Comorbidity of reading and mathematics disabilities: Genetic and environmental etiologies. *Journal of Learning Disabilities, 28*, 96–106.

Locuniak, M. N., & Jordan, N. C. (2008). Using kindergarten number sense to predict calculation fluency in second grade. *Journal of Learning Disabilities, 41*, 451–459.

Logie, R. H. (1995). *Visuo-spatial working memory*. Hove, UK: Erlbaum.

Lopez, M. F., & Forness, S. R. (1996). Children with attention deficit hyperactivity disorder or behavioral disorders in primary grades: Inappropriate placement in the learning disability category. *Education & Treatment of Children, 19*, 286–299.

Mandler, G., & Shebo, B. J. (1982). Subitizing: An analysis of its component processes. *Journal of Experimental Psychology: General, 111*, 1–22.

Mazzocco, M. M. M. (2007). Defining and differentiating mathematical learning disabilities and difficulties. In D. B. Berch & M. M. M. Mazzocco (Eds.), *Why is Math so Hard for Some Children? The Nature and Origins of Mathematical Learning Difficulties and Disabilities* (pp. 29–48). Baltimore, MD: Paul H. Brookes Publishing Co.

Mazzocco, M. M. M., & Devlin, K. T. (2008). Parts and 'holes': Gaps in rational number sense in children with vs. without mathematical learning disability. *Developmental Science, 11*, 681–691.

Mazzocco, M. M. M., & Kover, S. T. (2007). A longitudinal assessment of executive function skills and their association with math performance. *Child Neuropsychology, 13*, 18–45.

McLean, J. F., & Hitch, G. J. (1999). Working memory impairments in children with specific arithmetic learning difficulties. *Journal of Experimental Child Psychology, 74*, 240–260.

Murphy, M. M., Mazzocco, M. M. M., Hanich, L. B., & Early, M. C. (2007). Cognitive characteristics of children with mathematics learning disability (MLD) vary as a function of the cutoff criterion used to define MLD. *Journal of Learning Disabilities, 40*, 458–478.

National Mathematics Advisory Panel, (2008). *Foundations for success: Final report of the National Mathematics Advisory Panel*. Washington, DC: United States Department of Education. *http://www.ed.gov/about/bdscomm/list/mathpanel/report/final-report.pdf*.

Ohlsson, S., & Rees, E. (1991). The function of conceptual understanding in the learning of arithmetic procedures. *Cognition and Instruction, 8*, 103–179.

Oliver, B., Harlaar, N., Hayiou-Thomas, M. E., Kovas, Y., Walker, S. O., Petrill, S. A., *et al.* (2004). A twin study of teacher-reported mathematics performance and low performance in 7-year-olds. *Journal of Educational Psychology, 96*, 504–517.

Ostad, S. A. (1997). Developmental differences in addition strategies: A comparison of mathematically disabled and mathematically normal children. *British Journal of Educational Psychology, 67*, 345–357.

Ostad, S. A. (1998). Developmental differences in solving simple arithmetic word problems and simple number-fact problems: A comparison of mathematically normal and mathematically disabled children. *Mathematical Cognition, 4*, 1–19.

Paglin, M., & Rufolo, A. M. (1990). Heterogeneous human capital, occupational choice, and male–female earning differences. *Journal of Labor Economics, 8*, 123–144.

Passolunghi, M. C., Cornoldi, C., & De Liberto, S. (1999). Working memory and intrusions of irrelevant information in a group of specific poor problem solvers. *Memory & Cognition, 27*, 779–790.

Passolunghi, M. C., & Siegel, L. S. (2004). Working memory and access to numerical information in children with disability in mathematics. *Journal of Experimental Child Psychology, 88*, 348–367.

Passolunghi, M. C., Vercelloni, B., & Schadee, H. (2007). The precursors of mathematics learning: Working memory, phonological ability and numerical competence. *Cognitive Development, 22*, 165–184.

Posner, M. I., Boies, S. J., Eichelman, W. H., & Taylor, R. L. (1969). Retention of visual and name codes of single letters. *Journal of Experimental Psychology Monograph, 79*, 1–16.

Raghubar, K., Cirino, P., Barnes, M., Ewing-Cobbs, L., Fletcher, J., & Fuchs, L. (2009). Errors in multi-digit arithmetic and behavioral inattention in children with math difficulties. *Journal of Learning Disabilities, 42*, 356–371.

Rivera-Batiz, F. (1992). Quantitative literacy and the likelihood of employment among young adults in the United States. *Journal of Human Resources, 27*, 313–328.

Rourke, B. P. (1993). Arithmetic disabilities, specific and otherwise: A neuropsychological perspective. *Journal of Learning Disabilities, 26*, 214–226.

Russell, R. L., & Ginsburg, H. P. (1984). Cognitive analysis of children's mathematical difficulties. *Cognition and Instruction, 1*, 217–244.

Shalev, R. S., Manor, O., & Gross-Tsur, V. (2005). Developmental dyscalculia: A prospective six-year follow-up. *Developmental Medicine & Child Neurology, 47*, 121–125.

Shalev, R. S., Manor, O., Kerem, B., Ayali, M., Badichi, N., Friedlander, Y., et al. (2001). Developmental dyscalculia is a familial learning disability. *Journal of Learning Disabilities, 34*, 59–65.

Siegel, L. S., & Ryan, E. B. (1989). The development of working memory in normally achieving and subtypes of learning disabled children. *Child Development, 60*, 973–980.

Siegler, R. S. (1996). *Emerging minds: The process of change in children's thinking.* New York: Oxford University Press.

Siegler, R. S., & Booth, J. L. (2004). Development of numerical estimation in young children. *Child Development, 75*, 428–444.

Siegler, R. S., & Jenkins, E. (1989). *How children discover new strategies.* Hillsdale, NJ: Erlbaum.

Siegler, R. S., & Robinson, M. (1982). The development of numerical understandings. In H. Reese & L. P. Lipsitt (Eds.), *Advances in Child Development and Behavior* (Vol. 16, pp. 241–312). New York: Academic Press.

Siegler, R. S., & Shrager, J. (1984). Strategy choice in addition and subtraction: How do children know what to do? In C. Sophian (Ed.), *Origins of Cognitive Skills* (pp. 229–293). Hillsdale, NJ: Erlbaum.

Starkey, P. (1992). The early development of numerical reasoning. *Cognition, 43*, 93–126.

Starkey, P., & Cooper, R. G.Jr., (1980). Perception of numbers by human infants. *Science, 210*, 1033–1035.

Strauss, M. S., & Curtis, L. E. (1984). Development of numerical concepts in infancy. In C. Sophian (Ed.), *Origins of Cognitive Skills: The Eighteenth Annual Carnegie Symposium on Cognition* (pp. 131–155). Hillsdale, NJ: Erlbaum.

Swanson, H. L. (1993). Working memory in learning disability subgroups. *Journal of Experimental Child Psychology, 56*, 87–114.

Swanson, H. L., Jerman, O., & Zheng, X. (2008). Growth in working memory and mathematical problem solving in children at risk and not at risk for serious math difficulties. *Journal of Educational Psychology, 100*, 343–379.

Swanson, H. L., Jerman, O., & Zheng, X. (2009). Math disabilities and reading disabilities: Can they be separated? *Journal of Psychoeducational Assessment, 27*, 175–196.

Swanson, H. L., & Malone, S. (1992). Social skills and learning disabilities: A meta-analysis of the literature. *School Psychology Review, 21*, 427–441.

Swanson, H. L., & Sachse-Lee, C. (2001). Mathematical problem solving and working memory in children with learning disabilities: Both executive and phonological processes are important. *Journal of Experimental Child Psychology, 79,* 294–321.

Swanson, H. L., Trainin, G., Necoechea, D. M., & Hammill, D. D. (2003). Rapid naming, phonological awareness and reading: A meta-analysis of the correlation evidence. *Review of Educational Research, 73,* 407–440.

Taub, G. E., Floyd, R. G., Keith, T. Z., & McGrew, K. S. (2008). Effects of general and broad cognitive abilities on mathematics achievement. *School Psychology Quarterly, 23,* 187–198.

Temple, C. M. (1991). Procedural dyscalculia and number fact dyscalculia: Double dissociation in developmental dyscalculia. *Cognitive Neuropsychology, 8,* 155–176.

Thorell, L. B., Lindqvist, S., Nutley, B. S., Bohlin, G., & Klingberg, T. (2009). Training and transfer effects of executive functions in preschool children. *Developmental Science, 12,* 106–113.

Walberg, H. J. (1984). Improving the productivity of America's schools. *Educational Leadership, 41,* 19–27.

Wynn, K. (1992). Addition and subtraction by human infants. *Nature, 358,* 749–750.

Wynn, K., Bloom, P., & Chiang, W.-C. (2002). Enumeration of collective entities by 5-month-old infants. *Cognition, 83,* B55–B62.

THE POOR COMPREHENDER PROFILE: UNDERSTANDING AND SUPPORTING INDIVIDUALS WHO HAVE DIFFICULTIES EXTRACTING MEANING FROM TEXT

Paula J. Clarke, * *Lisa M. Henderson,*[†] *and Emma Truelove*[†]

* SCHOOL OF EDUCATION, UNIVERSITY OF LEEDS, LEEDS, UNITED KINGDOM
[†] DEPARTMENT OF PSYCHOLOGY, UNIVERSITY OF YORK, HESLINGTON, YORK, UNITED KINGDOM

Advances in Child Development and Behavior
Patricia Bauer : Editor

I. Introduction

Extracting meaning from written language is an extremely important process, which facilitates communication, learning, and everyday functioning. Sometimes this process can seem relatively automatic but at other times it may be highly effortful. It is known that many factors play a role in determining the ease with which written language is understood.

Comprehending text involves highly complex skills that can place heavy demands on cognitive and metacognitive resources. Beyond the mechanics of reading words, a solid understanding of word meanings is essential; this helps the reader to build a literal interpretation of text. This process, however, is not always straightforward; many words have multiple meanings and individuals need to be able to use sentence context and grammar cues to choose which meaning is appropriate. In addition, text is not always intended to be interpreted literally; passages often include examples of figurative language such as simile, metaphor, and idiom that require the reader to bring prior knowledge of phrases and expressions to the text. To complicate the comprehension process further, text rarely provides the reader with all the details needed to understand it fully; inferences frequently need to be made.

This chapter provides an overview of the challenges associated with reading for meaning; it will introduce key theoretical models and different reading profiles. The main focus of this chapter will be the "poor comprehender profile" which refers to individuals who despite being able to read accurately experience difficulties in extracting meaning from text. To fully understand the nature of this profile, attention needs to be paid to multiple levels of explanation. According to the Morton and Frith (1995) causal framework for understanding developmental disorders, factors must be considered at behavioral, cognitive, biological, and environmental levels with an appreciation of the relationships and interactions between these levels. This chapter aims to provide an overview of current knowledge at each of these levels; first, the behavioral features of the poor comprehender profile will be examined with particular focus on the criteria and assessments used to quantify reading comprehension. Following this, we shall present a summary of cognitive explanations; specifically language (semantics, grammar, pragmatics, inferencing), working memory, inhibition, and metacognition. Factors associated with the poor comprehender profile at the biological level will then be introduced before considering environmental factors and the impact of interventions. Key issues facing research and practice in this field will be highlighted throughout and avenues of promising development identified.

A. KEY THEORIES OF READING COMPREHENSION

In this section, we introduce two theories that have been used to describe and account for factors relating to the poor comprehender profile; the "simple view of reading" (Gough & Tunmer, 1986) and the Construction-Integration model (CI model; Kintsch & Rawson, 2005). Our aim is not to provide an exhaustive review of theories of reading, rather we have chosen to focus on the two that we refer to most frequently in the main body of this chapter.

In 1986, Gough and Tunmer proposed the "simple view of reading" to encapsulate the skills necessary for reading comprehension. The simple view comprises two components, decoding and linguistic comprehension, and argues that both are necessary for understanding text. Crucially, it acknowledges that neither component alone is sufficient for reading comprehension success and that the two components have a multiplicative rather than additive relationship. Hoover and Gough (1990) clarify the terms used for these components by stating that "decoding" is used to encompass all aspects of efficient word recognition and "linguistic comprehension" refers to the ability to use lexical information to derive sentence and discourse interpretations. This model has been incredibly influential in research over the past two decades; in particular, it has been used extensively to investigate different reading profiles. Before exploring this, however, it is important to acknowledge that there is variability in the literature regarding the exact terminology used to convey the simple view. The terms "linguistic comprehension," "language comprehension," and "listening comprehension" have all been used in published descriptions. For the purposes of this chapter, "listening comprehension" will be used; this most accurately relates to the research methods that have been used to test the model, and is more specific than the terms "language comprehension" and "linguistic comprehension" which could be misinterpreted to mean understanding of both written and spoken language.

The simple view can be used to predict four different reading outcomes (see Figure 1). This application of the theory has been referred to as the Reading Component Model (Catts, Hogan, & Fey, 2003). The first outcome is successful reading in which both decoding and listening comprehension components are intact. The second is the opposite to this, a generally poor reading profile in which there is evidence of impairments in both decoding and listening comprehension components.

The third is the dyslexic profile in which decoding is impaired but listening comprehension is preserved. Precise definitions of dyslexia vary. Historically they have implied a discrepancy between general cognitive ability and reading skills, for example, the Orton Dyslexia Society

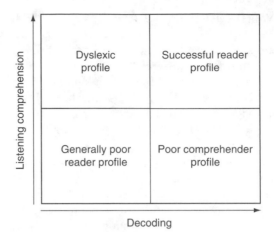

Fig. 1. The reading profiles predicted by the simple view of reading (Gough & Tunmer, 1986).

(1994) recognize difficulties in single word reading that "... are often unexpected in relation to age or other cognitive abilities; they are not the result of generalized developmental disability or sensory impairment" (cited in Snowling, 2000, p. 24). More recently, Hulme and Snowling (2009) have defined dyslexia as "a problem in learning to recognize printed words at a level appropriate for a child's age" (p. 39), removing the notion that in order to meet the criteria for dyslexia other cognitive abilities need to be intact. An independent review by Rose (2009), for the Department for Children, Schools and Families (DCSF), defines dyslexia as "a learning difficulty that primarily affects the skills involved in accurate and fluent word reading and spelling" (p. 10). In line with Hulme and Snowling, the report continues by stating, "Dyslexia occurs across the range of intellectual abilities" (p. 10) and "It is best thought of as a continuum, not a distinct category, and there are no clear cut-off points" (p. 10).

The final profile, the focus of this chapter, is the poor comprehender profile characterized by listening comprehension difficulties with intact decoding. The poor comprehender profile has been described in a variety of different ways, with the terms "poor comprehender," "less-skilled comprehender," "specific comprehension deficit," "specific reading comprehension impairment," and "hyperlexia" all being used in the research literature. Originally outlined by Oakhill (1981), this profile describes the significant minority of children who experience difficulty making the transition from learning to read to reading to learn. Poor comprehenders present the mirror image of dyslexia: they have difficulty in understanding what they read, despite having age-appropriate word reading accuracy and fluency, and normal phonological decoding skills (Cain & Oakhill, 2009; Nation, 2009). The specific criterion used to classify poor comprehenders varies considerably across studies and the implications of this will be discussed in a later section of this chapter. It is important to make the distinction between the terms "poor comprehender" and "hyperlexia" as there is sufficient reason to view these two profiles as separable. Nation (1999) conducted a review of the literature and suggested that individuals who show a hyperlexic profile often have cognitive, language, and social impairments alongside word recognition skills that are well in advance of reading comprehension ability. The review goes on to pinpoint early onset of word recognition ability and preoccupation with reading as two further features of hyperlexia. Although the poor comprehender profile has a discrepancy between reading accuracy and comprehension at its core, usually poor comprehenders have general cognitive ability in the average range and factors such as early onset of word recognition and obsessive reading behaviors are not part of selection criteria. The poor comprehender profile is not yet widely recognized by practitioners and it has received considerably less attention than dyslexia in the research literature. This chapter aims to raise awareness of this profile and explore the extent to which it represents a discrete category of reading behavior. Throughout this chapter, for consistency and clarity the term "poor comprehender" will be used.

Catts, Adlof, and Weismer (2006) promote the use of the simple view in categorizing reading profiles and argue that discrepancies in performance on its two components should be used to distinguish between individuals with dyslexia and poor comprehenders. They conducted two studies, the findings of which corroborate this advice. The first showed that on measures of receptive vocabulary, discourse comprehension, and discourse inferencing, poor comprehenders performed significantly worse than poor decoders and typical readers, whereas, on measures of phonological processing, the poor decoders group performed significantly worse than other groups. The second retrospectively examined the performance of the three groups at earlier time points (second and fourth grades) and showed that patterns of language comprehension and phonological deficits were consistent over time. Catts *et al.* suggest that the stability of the wider language profiles, derived from the simple view, can inform the early identification and intervention of children at risk of developing reading disorders.

The simple view is a compelling but ultimately reductionist theory that can be used in a very broad way to capture the nature of reading comprehension (see Hoffman, 2009, for a critique). Its simplicity does not capture the complex nature of comprehension which involves numerous processes and cognitive skills. Modifications to the model have been proposed, for example, Hulme and Snowling (2009) have speculated that the simple view could be

extended by including factors such as comprehension monitoring and motivation to read. Furthermore, Joshi and Aaron (2000) have argued that the
simple view could be adapted to include a speed of processing component.
They found that rapid automatized naming (RAN) accounted for an additional 10% of unique variance in reading comprehension above and beyond
measures of decoding and listening comprehension. They suggested that an
additive function of speed of processing could be linked to the multiplicative
relationship between decoding and listening comprehension in the simple
view. In addition, Vellutino, Tunmer, Jaccard, and Chen (2007) have highlighted the need to consider the changing relationships between decoding
and listening comprehension over time and have argued that the model
needs to account for the different contributions of each component to
reading comprehension across development.

Kintsch and colleagues have proposed a psycholinguistic framework for
understanding the processes involved in reading comprehension, the CI
model (Kintsch & Rawson, 2005). This framework acknowledges the complexities of reading comprehension and views the processes involved as
heavily interactive. Kintsch and Rawson describe comprehension as
involving processing at different levels; these are outlined in Figure 2.

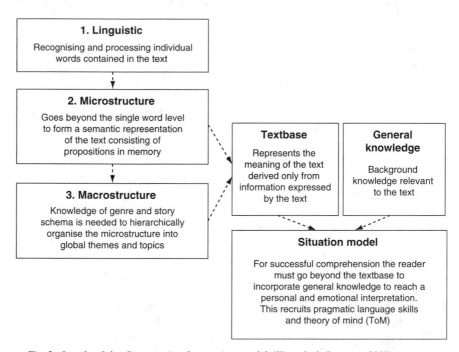

Fig. 2. Levels of the Construction Integration model (Kintsch & Rawson, 2005).

Importantly, a distinction is made between the contributions of the text itself and the reader's prior knowledge. The authors recognize that "... if a reader only comprehends what is explicitly expressed in a text, comprehension will be shallow, sufficient perhaps to reproduce the text, but not for a deeper understanding" (Kintsch & Rawson, 2005, p. 211). The model proposes that comprehension requires efficient processing of text (termed the text base which has three components, linguistic processing, microstructure, and macrostructure) and successful activation and integration of what is termed the situation model. The situation model comprises background knowledge, personal and emotional motivations and perspectives, imagery, understanding of the pragmatic aspects of language, and theory of mind skills. Central to the success of the interactive processing between the text base and the situation model is the ability to make inferences and the authors suggest that "Resource demanding, controlled inference processes in text comprehension are not restricted to knowledge retrieval. The situation model for a literary text may require construction at more than one level of analysis; to understand a story; the reader may have to infer the protagonists' motivations; to understand an argument, the exact relations between its components may have to be analyzed" (Kintsch & Rawson, p. 221). The CI model views comprehending text as a highly complex network of interacting systems and processes; it prompts research to investigate the various cognitive skills and resources that feed into reading comprehension and provides a framework for considering the specific areas in which poor comprehenders may show impairments.

However, a limitation of the CI model is that it assumes that word recognition is fluent and there is automatic access to meaning at the level of the single word. Also as with the simple view, the CI model does not take a developmental perspective and future studies must endeavor to fit this model into a framework that can be applied to the developing reader.

II. Behavioral Profile

To understand the behavioral profile of poor comprehenders, it is necessary to first consider how behavior is measured. This section will therefore begin with an examination of how attainment in reading comprehension is quantified. Following this, attention will be paid to the definitions and criteria used in the research literature to describe and select poor comprehender samples. As the poor comprehender profile is not represented in formal diagnostic criteria, it is vital that differences between studies are recognized so that meaningful conclusions can be drawn across findings.

A. ISSUES OF ASSESSMENT

To categorize reading disorders using the simple view, it is imperative that valid and reliable psychometric tests are used. A concern in the identification of poor comprehenders is whether it is possible to quantify the extent to which text is understood. If we acknowledge that reading comprehension is inherently complex and involves a multitude of factors, then it is important to consider whether it can be measured using a single test. If we are to monitor reading comprehension over time then an assessment that provides a valid measurement at different points in development is required.

A recent report from an Economic and Social Research Council (ESRC, unpublished) seminar series on reading comprehension recognizes a further issue, "The complexity of reading comprehension presents challenges for assessment, especially as many of the cognitive processes that contribute to reading comprehension are covert and therefore cannot be directly observed or measured" (p. 3). This prompts consideration of confounds associated with the indirect procedures used in most reading comprehension measures.

The majority of standardized reading comprehension assessments use off-line methodologies that measure the end point of comprehension not the process itself. Commonly, readers are given a series of sentences or short passages to read and following each they are asked a set of questions to tap understanding. These tests vary in a number of important ways, most notably the types of reading material used (e.g., fictional narratives, expository texts), the information probed by the questions, the question types, and the response method. Different assessments tap different aspects of reading comprehension to varying degrees, and the prevalence of reading comprehension impairments and the nature of the poor comprehender profile vary according to the test used to assess comprehension (Bowyer-Crane & Snowling, 2005; Keenan, Betjemann, & Olson, 2008). Furthermore, passage reading accuracy (which involves fluent word identification and decoding) is inextricably linked to reading comprehension, and as such it is not possible to create a test of reading comprehension that is not influenced to some extent by decoding ability; decoding will always contribute variance to reading comprehension skill. The Neale Analysis of Reading Ability II (NARA-II, Neale, 1997) and the primary version of the York Assessment of Reading for Comprehension (Snowling *et al.*, 2009) attempt to overcome the effects of passage reading accuracy errors on reading comprehension by correcting reading errors and imposing a discontinuation rule when too many errors have been made. However, Spooner, Baddeley, and Gathercole (2004) argue that the scores from the NARA-II should be interpreted with caution as they can underestimate the comprehension ability of children with weak decoding skills. It follows that to gain a true picture of an individual's reading comprehension skills it is important to conduct separate measures of its component skills (decoding and listening comprehension) and reading comprehension.

Bowyer-Crane and Snowling (2005) draw a distinction between tests in terms of method of reading (aloud or silently). On tests such as the Wechsler Objective Reading Dimensions (WORD, Wechsler, 1990) Reading Comprehension subtest which is read silently or the Wechsler Individual Achievement Test II (WIAT-II, Wechsler, 2001) which can be read silently or aloud, it is difficult to assess the extent to which an individual's comprehension might be disrupted by poor reading accuracy. In some circumstances, however, it may be preferable to use a silent reading method, for example, secondary school aged readers rarely read aloud. For this reason, the YARC secondary version (Snowling *et al.*, 2010) uses a silent reading procedure.

The type of reading material influences the range of comprehension questions that can be asked and consequently how an individual's performance should be interpreted. Bowyer-Crane and Snowling (2005) compared form 1 of the NARA-II (Neale, 1997) and the WORD Reading Comprehension subtest (Wechsler, 1990). The NARA-II uses passages that largely follow a narrative structure whereas the WORD uses a more expository style. Bowyer-Crane and Snowling qualitatively analyzed the information probed by the questions used in these tests. They found that 32% of the questions in the WORD could be answered using literal information compared to 14% of the questions from the NARA-II. They also observed the NARA-II involved more cohesive and knowledge based inferencing whereas the WORD prompted more elaborative inferencing. Moreover, poor comprehenders identified using the NARA-II performed in the average range on the WORD further highlighting the differences between the two tests.

An additional consideration is the extent to which the content of the reading material is consistent with an individual's own experience. Bowyer-Crane and Snowling (2008) used an online verification task to ascertain whether or not 10-year-olds have more difficulty generating inferences when they are based on fictional information that runs contrary to their knowledge of the world. Children were asked to read a story and make true/false judgments about critical test sentences representing either a causal (coherence) or a static (elaborative) inference in either Real World or Fairy Story context. For both contexts, faster and more accurate judgments were made in the causal inference condition than in the static inference condition. However, judgments made to critical test sentences were slower in the Fairy Stories. These findings suggest that inference generation may be harder for children when they are reading fictional stories, arguably because they require the suspension of disbelief. Although yet to be specifically examined, it is possible that poor comprehenders have more difficulties reading fictional passages when there is an increased demand on inference generation.

These findings have important implications for assessment, suggesting that a variety of passages (including literal and nonliteral information, fiction and nonfiction) should be administered to assess and identify the poor comprehender profile. Closely related to this issue is the importance of selecting age- and gender-appropriate passages that children are motivated to comprehend (Guthrie & Wigfield, 2000).

A somewhat controversial yet seldom formally debated issue is whether the text should remain in front of the child as they answer comprehension questions. If the text is removed, there is a significantly higher memory load on comprehension because children are denied the opportunity to reread relevant parts of the text. Taking the text away possibly underestimates a child's comprehension ability; when selecting poor comprehenders it may also bias recruitment toward children who have memory and attention problems. It is arguable that the prevalence of poor comprehenders may be higher and the population may be more heterogeneous when the text is removed. De Beni and Palladino (2000) explicitly stated that their procedure allowed participants to look back to the passages during the questions so that a purer measure of reading comprehension could be obtained and "... participants could be selected who differed in comprehension abilities as independently as possible from long-term memory capacity" (p. 134). Researchers should adopt a common practice to classify poor comprehenders and clearly document why one approach has been chosen over the other.

Tests use different types of questions and response methods to gauge understanding. Most individually administered tests require a verbal response to an open-ended question. This is problematic for a number of reasons, first, the participant must be able to understand what is being asked of them, otherwise the test becomes as much about question comprehension as comprehension of the reading material itself. Second, the testing situation requires a degree of social interaction and pragmatic language understanding, the participant needs an awareness of how much information to provide, and this can be influenced by the assessor's tone of voice, prompting behaviors, body language, eye contact, etc. Third, the participant must provide a coherent response that is relevant and contains sufficient detail to fully answer the question. This might be challenging for some individuals, for example, those who have expressive language difficulties, problems with organization and planning, working memory impairments, shyness, etc. The spoken response may not truly reflect the individual's level of understanding. Finally, the assessor is required to quantify the quality of the response using a predefined scoring system and subjectivity or bias can be unintentionally introduced into the scoring process. Spooner *et al.* (2004) recognize that much of the research on poor comprehenders has used the NARA-II, which employs a verbal response procedure to identify comprehension groups and suggests that "Such research may be reporting circular effects of ability to answer open-ended questions" (p. 201). They go on to suggest that assessments that use this methodology should be avoided in comprehension research.

An alternative type of question format is multiple choice; this is used in the Gray Oral Reading Test 4 (GORT-4, Wiederholt & Bryant, 2001). Although useful for children with expressive language weaknesses, Keenan and Betjemann (2006) highlight that passage-independent questions are particularly problematic for multiple choice formats. They represent a serious confound as they are susceptible to guessing and above chance levels of responding. Keenan and Betjemann found that undergraduates were able to answer 86% of the questions correctly more than 25% of the time (chance level responding). Care must therefore be taken to ensure that the content validity of tests is scrutinized to ensure results are not misinterpreted. In summary, there are many factors which influence performance on tests of reading comprehension. To obtain the most accurate estimate of ability it is advisable to employ multiple measures; varying the type of reading material, how the reading material is read (silently or aloud), the question types and the response method.

B. CRITERIA

Research studies of poor comprehenders vary immensely in the criteria and the assessment procedures that they use to define the poor comprehender profile. This has important implications for testing hypotheses and making comparisons between research studies.

The extent to which the reading comprehension of poor comprehenders deviates from the level expected for their age varies across studies. Some have used a cutoff of reading comprehension being at least 1 year below chronological age (Nation & Snowling, 1998a, 1999; Cain & Oakhill, 2006; Weekes, Hamilton, Oakhill, & Holliday, 2008; Cain, 2006) or more than 10 or 15 standard score points below the test mean of 100 (Nation, Marshall, & Altmann, 2003; Catts *et al.* (2006); Ricketts,

Bishop, & Nation, 2008; Pimperton & Nation, 2010). Others have not specified exactly how much reading comprehension ability should deviate from chronological age (Cain & Oakhill, 1999; Cain, Oakhill, Barnes, & Bryant, 2001; Cain, Oakhill, & Lemmon, 2004, 2005).

There is also a lack of consensus as to how impaired reading comprehension should be in relation to reading accuracy. Studies that take this into account have adopted differing criteria: (a) reading comprehension must be at least 6 months (Stothard & Hulme, 1995; Cain & Oakhill, 1999), 8 months (Cain & Oakhill, 2006; Cain *et al.*, 2004), or 1 year (Cain *et al.*, 2005; Cain, 2006; Nation & Snowling, 1998b; Nation *et al.*, 2003; Weekes *et al.*, 2008) below text reading accuracy age, or (b) reading comprehension must at least one standard deviation (15 standard score points) below text reading accuracy (Catts *et al.*, 2006). Across studies, these criteria give rise to a range of discrepancies between reading comprehension and reading accuracy; with averages varying from 12 to 32 months in age-equivalent scores and from 17 to 34 in standard score points.

Instead of anchoring reading comprehension against word or passage reading accuracy to identify poor comprehenders, Nation and colleagues have advocated the use of nonword decoding (e.g., Nation, Clarke, Marshall, & Durand, 2004; Nation & Snowling, 1998a, 1999; Ricketts *et al.*, 2008). Nation and Snowling (1997), following an investigation of the performance of 184 children aged between 7 and 9 years on the NARA, reported that nonword decoding is largely independent of comprehension in contrast to text reading accuracy which is positively related to reading comprehension. Therefore, poor comprehenders may show poorer text reading accuracy than would be predicted on the basis of their basic-level decoding. To address this issue, Nation and Snowling (1998a) equated their poor comprehender and control groups for decoding skill (nonword reading) instead of text reading accuracy. To ensure that no generally poor readers were selected, all poor comprehenders obtained at least age-appropriate text reading accuracy.

In addition to ensuring that poor comprehenders have word reading and nonword decoding skills in the average range, Cain, Oakhill, and colleagues have also selected poor comprehenders who have verbal IQ in the normal range (Cain, 2006; Cain & Oakhill, 1999; Cain *et al.*, 2001, 2004, 2005). In these studies, measures of expressive and receptive vocabulary have often been used as a proxy for verbal IQ. The reasoning behind this approach is that by selecting poor comprehenders who have age-appropriate verbal ability you effectively factor out vocabulary knowledge (or access to relevant knowledge) as a cause of reading comprehension difficulties and this enables the investigation of higher-level comprehension skills such as inference making. This contrasts to other studies in which vocabulary knowledge has been free to vary (e.g., Nation & Snowling, 1998a, 1998b, 1999;

Nation *et al.*, 2003; Catts *et al.*, 2006; Cain & Oakhill, 2006; Ricketts *et al.*, 2008; Pimperton & Nation, 2010). This latter research has tended to be more concerned with investigating the nature of the vocabulary impairment. Nation *et al.* (2004) reported that weaknesses in oral language skills such as expressive vocabulary knowledge, morphosyntax, and communicative skills (including understanding communicative intent, lexical and syntactic ambiguity, and inference making) often characterize children with poor comprehender profiles. Arguably then, poor comprehenders who have been selected to have age-appropriate verbal skills may have different underlying cognitive profiles to poor comprehenders with weak verbal skills.

In summary, there has been considerable variation regarding how much reading comprehension should deviate from age-appropriate levels and reading accuracy or nonword decoding to be considered as impaired. It is crucial to put in place a working definition of the poor comprehender profile which enables direct comparison across research studies and promotes the advancement of theoretical understanding. Perhaps the most stringent criteria for defining reading comprehension impairment is a one standard deviation discrepancy in standard score points on measures of word reading accuracy and reading comprehension; where reading comprehension falls below the average range and text reading accuracy and nonword reading fall within the average range. Although such criteria is useful when selecting children for research studies, reading disorders may be best thought of as extremes of the normal distribution of reading ability rather than distinct categories. Furthermore, children that do not fit the stringent categorical approach may still benefit from intervention designed to boost comprehension skills.

C. BEHAVIORAL DIFFICULTIES IN POOR COMPREHENDERS

An area that has yet to receive much research attention is the possibility that behavioral difficulties may contribute to or emerge as a result of the poor comprehender profile. For some children, behavioral difficulties may mask underlying reading comprehension difficulties and educational needs in the classroom. It is important that future research examines the behavioral profile of these children to ascertain the presence or absence of any associated behavioral difficulties. A promising area for investigation is the prevalence of symptoms of inattention and/or hyperactivity in poor comprehenders. It is acknowledged that effortful processing necessitates the ability to sustain attention over time (Hasher & Zacks, 1979). Given the effortful nature of reading comprehension and the demands placed on sustained attention, it follows that children with attention weaknesses may struggle to comprehend, especially when reading longer texts.

A small body of research has explored the reading comprehension skills of children with ADHD and there is evidence to suggest that some children with inattentive profiles demonstrate weaknesses in understanding what they read in the face of adequate decoding skills. Brock and Knapp (1996) examined the reading comprehension skills of children with primarily inattentive symptoms of ADHD who were preselected to have age-appropriate word reading ability. They found that these children displayed poorer reading comprehension skills than peers without ADHD who were matched for letter–word identification, age, and gender. In a review of studies investigating story comprehension in children with ADHD, Pugzles-Lorch, Milich, and Sanchez (1998) support the assertion that these children exhibit weaknesses in their understanding of stories across the course of development. To date there is only limited research into the reading comprehension skills of children with ADHD, and the extent to which poor comprehenders exhibit inattentive or hyperactive behavioral profiles is yet to be investigated. Nonetheless, it remains an interesting avenue to be explored, particularly across development as poor comprehenders become increasingly aware of their difficulties and the processes involved in reading comprehension make greater demands on attentional capacity. Moreover, evidence of any associated behavioral difficulties may have wide-ranging implications for intervention and be of considerable educational significance.

III. Cognitive Profile

The majority of research into the poor comprehender profile has focused on identifying core cognitive explanations. As reading comprehension is inherently complex it is reasonable to suggest that there may be multiple cognitive factors that contribute to the manifestation of the poor comprehender profile. The following section outlines current understanding of cognitive impairments in the areas of language, working memory, inhibition, and metacogniton and discusses measurement techniques that have advanced knowledge in this area and gaps in present knowledge.

A. LANGUAGE

1. Semantics

a. Access to semantic information. It has been estimated that a reader must know at least 90% of the words in a text in order to fully comprehend it (Nagy & Scott, 2000). In line with this view, Kintsch & Rawson (2005) specify that successful comprehension of written or spoken language first requires activating individual word meanings before integrating these meanings to build a coherent mental representation (see Figure 2). If word meanings are poorly represented in semantic memory, less information will be accessed and fewer links between concepts will be activated than if a rich semantic representation for word meaning exists.

One view is that the size or richness of an individual's vocabulary or the speed of access to vocabulary items affects reading comprehension ability. This view is central to the Perfetti (2007) lexical quality hypothesis, which suggests that word-level knowledge has direct consequences for comprehension. High lexical quality is defined as well-specified orthographic and phonological representations and clearly defined but flexible representations of meaning, allowing for quick and reliable meaning retrieval. An individuals' vocabulary includes words of widely varying lexical quality, from rare words to frequently occurring and well-known words. Individuals also differ in the average lexical quality of their words, which is independent of overall vocabulary size. Perfetti proposes that poor comprehenders with adequate orthographic and phonological skills may have word-level semantic problems that affect comprehension.

In line with this, there is evidence that poor comprehenders have impoverished vocabulary knowledge from which to draw on during comprehension. Poor comprehenders have weaker receptive and expressive vocabulary on standardized tests (e.g., Nation *et al.*, 2004). They perform worse on synonym judgment tasks (i.e., deciding whether word pairs have similar meanings, particularly when the words are more abstract) but not rhyme judgment tasks (i.e., deciding whether word pairs rhyme) compared to controls (Nation & Snowling, 1998a). This suggests that poor comprehenders have intact phonological skills but impairments within the semantic domain. Moreover, poor comprehenders are slower and less accurate at naming pictures than controls, particularly pictures with low frequency names (Nation, Marshall, & Snowling, 2001). They are also weaker than controls at producing as many exemplars as possible in response to semantic categories such as "animals" in a prespecified length of time (Nation & Snowling, 1998a). Together, these findings suggest that poor comprehenders have difficulties within the semantic domain, showing impoverished vocabulary knowledge particularly for more abstract words.

Many of the above studies are limited by the "off-line" nature of the tasks used to measure semantic processing. Off-line tasks such as meaning judgment, synonym judgment, and providing definitions do not reveal how the comprehension process unfolds over time but instead measure only the end point of processing. As argued by Swinney and Prather (1989), "if a task requires conscious introspection/analysis about the lexical process ... it will, at best, be a distant and impoverished reflection of this targeted process" (p. 228). In comparison, more dynamic "online" methodologies permit tracking behavior at a fine temporal resolution as processing unfolds over time. Such methodologies include eye tracking, event related potentials (ERPs), reading time paradigms and semantic priming. In recent years, there has been growing recognition of the importance of online methodologies to the field of developmental cognitive psychology and psycholinguistics; however, such methodologies are yet to be fully explored with poor comprehenders.

The semantic priming paradigm, introduced by Meyer and Schvaneveldt (1971), has been used to gain insight into how semantic context is used to access a word's meaning and to shed light on how semantic memory is organized and accessed (McNamara, 2005; Neely, 1991; Swinney, 1979; Taft, 1991). In this paradigm, a semantic context is provided by an auditory or visual word prime and is typically followed by a target word or picture. Responses to targets (e.g., CAT) are significantly faster and more accurate when they are preceded by prime words to which they are semantically related (e.g., DOG) than when they are unrelated (e.g., BOAT). This is known as the semantic priming effect.

Using this paradigm, Nation and Snowling (1999) found that the semantic deficit in poor comprehenders reflects weaker activation of abstract semantic associations between words. Children made lexical decisions toward aurally presented word primes and targets that were related or unrelated to each other (using an ISI of 500 ms). The poor comprehenders showed semantic priming for categorically related items that were highly associated (e.g., salt–pepper), for functional relationships that were associated (e.g., hammer–nail) and for functional relationships that were not associated (e.g., nose–head). However, they did not show priming for categorically related items that were not commonly associated (e.g., bed–desk). In contrast, controls (matched for age, decoding, and nonverbal ability) showed priming for all pair types. Associative and functional relationships are reinforced by lexical and real world co-occurrence whereas categorical relationships emerge out of increasingly refined knowledge of the semantic features of words. Thus, for poor comprehenders, semantic priming may be driven by lexical co-occurrence and not by semantic activation, and the connections between categorically related items may be weaker. Unfortunately, this study only used a single 500 ms ISI, and, because written presentation was used, the length of the ISI varied according to how fast the children responded to each word. Therefore, it remains possible that poor comprehenders show a different time course of semantic priming effects from controls rather than an absence of priming for certain relations.

The measurement of ERPs with either electroencephalography (EEG) or magnetoencephalography (MEG) is also critically important for the study of language. These methodologies provide a continuous temporal record of brain activity from the beginning of stimulus onset without interference from overt decision or response processes (Osterhout, 2000). Landi and Perfetti (2007) investigated ERPs that occur in response to semantic and phonological tasks in adults with poor comprehender profiles and adults who are skilled comprehenders. Participants saw a series of two pictures or words (presented one at a time) and were asked to decide if they were semantically related (the semantic task) or they heard two words (e.g., bear–bare) and were asked to decide if they sounded the same (the phonological task). For both tasks, an early positive going waveform was observed that began around 150 ms and continued until 250 ms (P200) and a negative going waveform was observed at around 400 ms (N400). Both P200 and N400 components were sensitive to differences between groups during the semantic (but not the phonological) task. These results suggest that individual differences in comprehension skill relate to neurophysiological markers of semantic but not phonological processing, supporting the simple view (Gough & Tunmer, 1986). Since Landi and Perfetti focused on adults, it remains to be determined whether neurophysiological markers of semantic processing relate to individual differences in comprehension skill in children.

Future studies using online methodologies including eye tracking and ERP in poor comprehenders will be important in helping to elucidate the time course of semantic processing and clarify the nature of the semantic obstacles to comprehension.

b. Acquiring new semantic information. It has been proposed that the ability to *acquire* new semantic information from context is the skill that mediates the relation between vocabulary and reading comprehension (Nippold, 2002; Sternberg & Powell, 1983). For instance, Nation, Snowling, and Clarke (2007) found that poor comprehenders have difficulty in learning semantic information about novel objects compared to controls (matched for nonword decoding, age, and nonverbal ability). Twelve poor comprehenders and 12 normal readers of the same age learned the names and meanings of four 3-syllable nonwords that had four semantic attributes (e.g., "*a corbealyon is a small, hairy, angry bird*"). A picture associated with each nonword was presented during the learning phase. Once they had learned the words and their meanings, the children completed two tasks: (i) they defined the words and (ii) they provided the words on hearing their definitions. The next day the children were asked to recall the phonological forms before they matched pictures to the newly learned words. The poor comprehenders took a similar number of trials to learn the new words as controls suggesting the phonological component of vocabulary learning is intact, but they had significantly more difficulty in defining the new words. There was no group difference in the immediate recall of the nonwords to definitions but after a delay of one day the poor comprehenders could remember fewer of the semantic attributes of the new words. These findings suggest that poor comprehenders have difficulty in learning (and consolidating) the semantic attributes of new words.

Similar findings have been obtained by Ricketts *et al.* (2008) in a nonword learning study with poor comprehenders and controls matched for age, nonverbal ability, and decoding skill. Children were trained to pronounce 20 visually presented nonwords (10 had consistent pronunciations and 10 had inconsistent pronunciations) with the opportunity to infer the meanings of the new words from a story context. Consistent nonwords were easier to learn than inconsistent words and both groups showed improved performance when asked to spell and make orthographic decisions about the nonwords after learning. Immediately after learning the nonword meaning, the poor comprehenders showed similar semantic learning to controls (on a nonword−picture matching task), but this knowledge was poorly retained over time. This study concurs with the studies by Nation *et al.* (2007), suggesting that poor comprehenders can make links between words and their semantic properties, but that they have difficulty in developing durable semantic representations.

This evidence suggests that as well as having poor existing oral vocabulary, poor comprehenders are worse at learning the meanings of new words. It could be argued that children who have smaller vocabularies (impaired breadth of vocabulary) or impoverished semantic representations (lower depth of vocabulary) from the outset may find it harder to learn new words because they have limited existing vocabulary to link with new representations. Shefelbine (1990) found that sixth grade students with the poorest vocabulary knowledge at the outset learned the fewest words from context, even though they had the greatest room for improvement. Therefore, any existing group differences in the quantity, quality, or organization of the semantic system may influence word learning.

c. Integration of semantic information. Once words have been recognized and their meanings retrieved, the meanings of the sentences (propositions) must be established before a text base and situation model can be formed (see Figure 2). Nation and Snowling (1998b) used the cross-modal priming task to investigate poor comprehenders' sensitivity to sentence context. The children heard low-constraining spoken sentences (e.g., "I went shopping with my mum and my ____") and then had to read contextually related or unrelated sentence-final target words that either had regular (e.g., "cash") or exceptional (e.g., "aunt") spelling-sound patterns. Children were also required to read the words in isolation. Poor comprehenders, children with dyslexia, and controls did not differ in accuracy when they read the words in isolation and all three groups benefited from the context (i.e., they showed better performance for related than unrelated conditions). However, the size of the facilitation effect correlated with listening comprehension skill and was smaller for the poor comprehenders than for the control and dyslexic groups. It was argued that semantic activation produced by the sentence primed the semantic pathway and facilitated word recognition to a greater extent for the dyslexic and control groups than for the poor comprehenders. However, because the sentences were constructed so that the final word was not predictable, it is likely that both grammatical and semantic knowledge contributed to performance in this task.

Using online methodology, Nation *et al.* (2003) provide conflicting evidence to the view that poor comprehenders show reduced sensitivity to context. Individual differences in children's online language processing was assessed by monitoring eye movements to objects in a visual scene as they listened to spoken sentences that contained biased or neutral verbs that were congruent or incongruent with the visual context (e.g., *Jane watched her mother choose the cake*, where all of the objects in the scene were "choosable," versus *Jane watched her mother eat the cake,* where the cake was the only edible object). Control children and poor comprehenders (matched for age, nonverbal ability, nonword decoding) made fast anticipatory eye movements to the target object (e.g., the cake), suggestive of similar sensitivity to linguistic constraints of the verb in each group. Thus, both groups were able to extract the semantic information from the verb within milliseconds and use it to guide ongoing processing by predicting subsequent reference to those objects in the visual scene that satisfied the verb's selection restrictions. Poor comprehenders activated this verb information as fast as control children. However, the poor comprehenders made more and shorter fixations to target objects, especially in the supportive condition. The authors speculated that this could be due to memory or attention deficiencies in the poor comprehenders, or problems with inhibiting the distracter pictures in the visual scene.

2. Grammar

Knowledge about syntactic constraints can aid the process of context integration because knowledge about the meanings and order of the noun and verb phrases may be insufficient to establish who did what to whom in sentences such as "The mouse that scared the elephant was chased by the cat." There are at least two ways in which grammatical knowledge may directly facilitate comprehension. First, grammatical understanding may facilitate the comprehension of sentences. Second, grammatical knowledge may facilitate the detection and correction of reading errors, thereby enhancing comprehension monitoring.

In the study by Catts *et al.* (2006) discussed earlier, poor comprehenders but not poor decoders showed deficits in grammatical skill, as well as impairments in vocabulary, discourse processing, and listening comprehension. Stothard and Hulme (1992) found that poor comprehenders showed weaker syntactic knowledge on the Test for Reception of Grammar (TROG; Bishop, 1982) which is a picture-sentence matching test. A similar result was obtained in a study by Cragg and Nation (2006) using the revised version of the TROG (TROG-2, Bishop, 2003) in which they found average standard scores of 80 and 94 for poor comprehenders and controls, respectively.

However, as with many other deficits shown in ability-matched studies of poor comprehenders, it is possible that poor grammatical understanding is a consequence, rather than cause, of poor comprehension. Gaining experience with syntactic structures that are uncommon in spoken language, such as the use of nominalizations and clausal noun phrases, may result from successful reading practice. Supporting this view in a longitudinal study of 7- to 9-year-olds, Oakhill, Cain, and Bryant (2003) reported that when verbal ability and vocabulary were controlled, syntactic ability was only a significant predictor of reading comprehension at the second of two test points, when children were older. This suggests the presence of grammatical impairments in poor comprehenders may depend on the stage of development.

However, a recent longitudinal study provides contradictory evidence to this view. Nation, Cocksey, Taylor, and Bishop (2010) identified 15 poor comprehenders (out of a total sample of 242 children) at 8 years of age and investigated their early language profiles prospectively at four time points from the point of school entry. Impairments in reading comprehension, but intact word reading accuracy and fluency, were persistent in the poor comprehenders over time. To investigate language profiles, the poor comprehenders were compared to 15 typically developing controls matched on reading accuracy and age. In support of the simple view (Gough & Tunmer, 1986), phonological skills were largely intact in the poor comprehenders but they showed consistent impairments in listening comprehension and grammar from 5 years of age. This suggests that such deficits precede rather than result from their reading comprehension impairments in mid-childhood. Notably, impairments in expressive vocabulary knowledge did not emerge until 6 years of age in the poor comprehender group, suggesting that difficulties with vocabulary knowledge become increasingly marked over time and are compounded by broader weaknesses in nonphonological oral language.

3. Pragmatics

In accordance with the CI model (see Figure 2), the ability to form a situation model of text or discourse is dependent on pragmatic language skill (Kintsch & Rawson, 2005). Pragmatic language refers to the social use of language including the ability to use social context in order to understand a speaker's or author's intention and to attain social goals through effective communication (Bishop, 1998).

It is yet to be determined whether poor comprehenders have weaknesses in pragmatic language; however, it is reasonable to hypothesize that comprehension difficulties may extend to problems with comprehending another's bid for conversation, including the comprehension of intended meaning. Nation *et al.* (2004) included a measure of communicative intent in their test battery and found that poor comprehenders are poorer than decoding-matched controls at planning and formulating oral expression. However, whether this finding was due to weaknesses in expressive language skills in the poor comprehenders, including semantics and syntax, or due to impairments in pragmatics is unclear.

If poor comprehenders have pragmatic difficulties when formulating verbal responses, this has implications for interpreting findings of poor performance on expressive vocabulary tests that require children to give definitions of words and, more importantly, for open-ended reading comprehension questions. The pragmatic demands of these assessment procedures, as outlined previously, might contribute to poor comprehenders' relatively weak performance. Studies are needed to investigate the nature of pragmatic language and the prevalence of pragmatic language impairment in poor comprehenders, and more generally, to further investigate the social and emotional profiles of these children. Indeed, the potential presence of pragmatic language impairments in poor comprehenders has important implications for the development of social skills and peer interaction (Lindsay, Dockrell, & Mackie, 2008).

4. Inferencing

Text and discourse rarely provide all of the details needed for successful interpretation. In order to construct an adequate and coherent representation (situation model), inferences need to be made by generating links between information provided and prior knowledge or between different parts of the text or discourse. Poor comprehenders have been shown to have problems with making inferences that depend on knowledge not explicitly stated in the text (Cain & Oakhill, 1999). For instance, Oakhill (1983) investigated the abilities of poor comprehenders to make instantiation inferences, where the reader has to infer a specific meaning of a common noun from the sentence context (e.g., inferring that "fish" is most likely a "shark" in the following sentence: "The fish frightened the swimmer"). Poor comprehenders made fewer instantiations than did skilled comprehenders, suggesting that their local processing of text is less influenced by the semantic content of the sentence. Poor comprehenders have also been shown to generate fewer constructive inferences that require the reader to integrate information from two different sources in the text (Oakhill, 1982). Furthermore, inference training has been shown to improve comprehension in small-scale studies of children with specific reading comprehension difficulties, these will be discussed later in this chapter (Yuill & Joscelyne, 1988; Yuill & Oakhill, 1988; McGee & Johnson, 2003).

As with deficits in grammatical understanding, satisfactory explanations for observed weaknesses in inference making in poor comprehenders are difficult to establish due to the dependence of inferences on lower-level comprehension processes and knowledge (Perfetti, Marron, & Foltz, 1996). As a result, there are a number of possible explanations for inference-making differences in poor comprehenders. One possibility is that deficits in vocabulary knowledge restrict poor comprehenders' inference making. Cain *et al.* (2001) investigated whether knowledge deficits are a possible cause of poor comprehenders' difficulties rather than the *process* of making an inference. Thirteen poor comprehenders and 13 good comprehenders (matched on age and reading accuracy) were exposed to information about an imaginary planet called "Gan" (e.g., "The flowers on Gan are hot like fire"). Once the children had learned the information they were read a multi-episode story and then answered questions to assess their inference-making ability. The inferences required the children to incorporate an item from the knowledge base with a premise in the text. Recall of the knowledge base was assessed at the end of the experiment and only responses to inference questions for which the knowledge base item was recalled were included in the data analysis. Even when knowledge was controlled for in this way, the poor comprehenders generated fewer inferences than the skilled comprehenders. These data suggest a problem with the process of inference making.

However, equivalent levels of performance on the knowledge questions do not mean that this knowledge was represented in memory in the same way in both groups of children. In fact, the poor comprehenders were slower to learn the new information and remembered less of this information over time. The poor comprehenders were also poorer at answering literal comprehension questions that did not require inference making. Therefore, this study cannot rule out that problems in inferencing do not reflect limitations in the storage, organization, or retrieval of semantic information.

The nature of the relationship between inference making and vocabulary acquisition is likely to be reciprocal. While poor vocabulary knowledge restricts the ability to make inferences, vocabulary acquisition often relies on the ability to make inferences from surrounding context. Evidence for this relationship is provided by Cain and colleagues who show that poor comprehenders are weak at inferring meanings of novel words in context (Cain *et al.*, 2004). Poor comprehenders were exposed to nonwords in short story contexts that provided cues for their meaning. The nonword was followed by a defining context either immediately or at a later point in the text. The children were required to "guess" the meaning of the nonword immediately (without contextual cues) and then again after completion of the story (with context). They found that poor comprehenders were less able to infer the meaning of the novel words from context than the skilled comprehenders and provided less rich definitions, particularly when the defining context appeared later in the text. An appreciation for the commonalities between learning new words from context and making inferences from text is important and has implications for training studies targeting these skills.

A second explanation for inference-making difficulties in poor comprehenders is that they lack the spontaneous drive to process language for meaning. Cain and Oakhill (1999) included a condition in which poor comprehenders were encouraged (with direct prompting) to search the text in order to find the information needed from which to make an inference. Interestingly, poor comprehenders' performance increased, leading Cain and Oakhill to suggest that it is not that poor comprehenders are unable to make inferences, but rather that they fail to do so spontaneously.

A third possibility is that poor comprehenders have processing limitations, such as poor working memory, which hamper their ability to make inferences and integrate text information with prior knowledge. Catts *et al.* (2006) reported that poor comprehenders were comparable to poor decoders and typical readers in making an inference when they remembered the premise on which it was based (when the premise was close in proximity to the point where an inference had to be made). However, they were significantly impaired relative to other groups when the premise and the inference were at a distance from each other in the text.

This suggests that poor comprehenders may set up their initial sentence representations adequately, but then subsequently fail to maintain or elaborate them.

B. WORKING MEMORY

Working memory refers to the simultaneous storage and processing of information (Baddeley, 2007). In a typical working memory task known as the listening span task, an individual may be asked to decide whether a series of sentences are true or false while retaining the sentence-final words. Language comprehension places heavy demands on working memory, and working memory skills are strongly correlated with reading comprehension (Cain *et al.*, 2004; Leather & Henry, 1994; Seigneuric & Ehrlich, 2005). Baddeley (1992) highlights two components from his model of working memory as being particularly necessary for reading comprehension; the phonological or articulatory loop (a temporary storage system for brief maintenance of verbal information) and the central executive (which oversees active manipulation of information in immediate memory and retrieval of information from long-term semantic memory).

Studies have shown that poor comprehenders perform poorly on working memory tasks. Yuill, Oakhill, and Parkin (1989) asked children to read aloud triplets of numbers and then to recall the final digit in each triplet. Poor comprehenders performed less well than control children, leading Yuill *et al.* to suggest that deficits in nonlinguistic working memory may underlie the reading comprehension problems seen in this group of children. However, it is arguable that these data are suggestive of a verbal memory deficit rather than a nonlinguistic deficit because the counting span task required children to read and recall digits both of which place demands on verbal processing.

Nation, Adams, Bowyer-Crane, and Snowling (1999) used Daneman and Carpenter's (1980) listening span task to investigate verbal working memory. This task's requirements, verifying short sentences and then recalling the last word of each sentence, are similar to some of the simultaneous processing and storage demands of language comprehension itself. Ten-year-old poor comprehenders showed listening span deficits compared to chronological age-matched controls, in line with the findings reported by Yuill *et al.* (1989). However, the groups performed similarly on a spatial working memory span measure. Similar findings were reported by Cain *et al.* (2004), who found that poor comprehenders showed significant difficulties on a version of the listening span task but performed normally on a counting span task (involving counting sets of dots and then recalling the numbers counted). These findings are consistent with the view that, rather than reflecting a general working memory impairment, poor comprehenders' weak performance on working memory tasks may be the consequence of impairments in verbal processing.

C. INHIBITION

It is likely that quick and efficient access to individual word meanings prepares the semantic system for later contextual processes. If a child fails to inhibit alternative meanings of words this could potentially interfere with the ability to form a coherent representation of a text or discourse. The intrusion of irrelevant semantic associations in the mental representation of the text would likely produce coherence breaks and heavily interfere with comprehension. Taking this view, Gernsbacher's structure building framework (Gernsbacher, 1993; Gernsbacher & Faust, 1991; Gernsbacher, Varner, & Faust, 1990) proposes that the ability to "suppress" irrelevant information varies with reading comprehension skill. Gernsbacher *et al.* (1990) found that typically developing adults and adults with poor reading comprehension skill were slower to decide that a target such as *ACE* was unrelated to *He dug with a spade* than *He dug with a shovel* at 100 ms stimulus onset asynchrony (SOA). At 850 ms SOA, only the poor comprehenders continued to show this interference effect. In a follow up experiment, Gernsbacher and Faust (1991) reported that both groups were similarly faster to accept that a target such as GARDEN was related to *He dug with the spade* than *He picked up the spade* at both short (100 ms) and long (1000 ms) SOAs. Therefore, it was argued that the adults with poorer reading comprehension had specific difficulties with inhibiting inappropriate information, but intact sensitivity to biasing sentence context. Similar findings have been obtained in children with hydrocephalus who also have poor comprehender profiles (Barnes, Faulkner, Wilkinson, & Dennis, 2004). However, the interpretation of these findings is complicated because the poor comprehenders in Gernsbacher's studies had concomitant word reading difficulties and the poor comprehenders in Barnes *et al.* were not matched to controls on nonverbal ability.

Studies have also reported that a difficulty with inhibiting "no longer relevant" information characterizes the verbal working memory performance of poor comprehenders (Cain, 2006; De Beni, Palladino, Pazzaglia, & Cornoldi, 1998; De Beni & Palladino, 2000; Palladino, Cornoldi, De Beni, & Pazzaglia, 2001). Poor comprehenders not only recall fewer items correctly in verbal working memory tasks, but they also make more intrusion errors than skilled comprehenders (recalling items from the experimental lists that have been processed but should have been suppressed).

Pimperton and Nation (2010) recently compared poor comprehenders and controls on verbal and nonverbal versions of a proactive interference task designed to assess the ability to suppress no longer-relevant information from working memory. The rate of intrusion of this to-be-forgotten verbal and nonverbal material was measured. The poor comprehenders showed domain-specific suppression deficits: They demonstrated higher rates of intrusion of material that they were instructed to forget than the controls, but only in the verbal version of the task. These findings are consistent with the view that, rather than reflecting a general difficulty with inhibition that is causal in their reading comprehension difficulties, poor comprehenders' difficulty with inhibition may be the consequence of impairments in accessing relevant semantic information.

An accumulating body of evidence suggests that poor comprehenders fail to inhibit irrelevant details when comprehending verbal material. At present, however, it is difficult to rule out the possibility that weaknesses in inhibition are due to lower-level difficulties in accessing relevant semantic information.

D. METACOGNITION

Comprehension monitoring refers to the metacognitive control processes that readers use to regulate their reading to the demands of the task and engage with the comprehension process. For example, if an individual realizes that their comprehension is inadequate, they can take appropriate steps to remedy the situation. There is evidence to suggest that poor comprehenders have weak metacognitive skills. For example, Yuill *et al.* (1989) presented scenarios containing anomalies, where resolving information was presented later in the scenario. Children were then asked comprehension questions about the anomalies, the rationale being that these questions can only be answered correctly if the different sources of information are integrated. Poor comprehenders were able to resolve anomalies when the pieces of information were close together in the text, but their performance suffered when the two pieces of information were further apart. Similar results were reported by Oakhill, Hartt, and Samols (1996). These results suggest that poor comprehenders may be able to monitor information adequately, but only when the task is made less demanding by decreasing the memory load. More recently, Oakhill *et al.* (2003) reported that performance on tests of comprehension monitoring at age 7 years predict reading comprehension ability a year later. Thus, there is preliminary evidence that this skill is necessary for the growth and development of comprehension skills.

A related concept is that of *standards of coherence*, originally proposed by van den Broek, Risden, and Husebye-Hartman (1995) and expressed as a working hypothesis by Perfetti, Landi, and Oakhill (2005) "... a standard of coherence broadly determines the extent to which a reader will read for understanding, make inferences, and monitor his or her own comprehension" (p. 233). Perfetti *et al.* argue that for reading comprehension to develop, a high standard of coherence is necessary. Standards of coherence are therefore an umbrella term that captures the extent to which an individual is concerned with whether text makes sense. Oakhill and Cain (2007) suggest that in order to establish whether a weak standard of coherence is a characteristic of the poor comprehender profile research attention should be paid to understanding the extent to which probable components of standards of coherence (e.g., comprehension monitoring and inferencing) co-occur in poor comprehenders. They raise the issue of whether it is possible to identify poor comprehenders who have impairment in one component but not the other, and explore possible avenues for the application of the concept of standards of coherence to intervention approaches.

Closely tied to monitoring and standards of coherence are motivational factors. Poor comprehenders may also fail to monitor text for inconsistencies because they lack the motivation or drive to read for meaning. Wigfield and Guthrie (1997) showed that motivation to read predicts growth in reading, and the amount of reading has been shown to be causally linked to gains in comprehension (Guthrie, Wigfield, Metsala, & Cox, 1999). At present it is unclear whether difficulties with comprehension monitoring and a lack of motivation to read for meaning are a cause or consequence of reading comprehension difficulties.

E. SUMMARY OF COGNITIVE PROFILE

There is a strong body of converging evidence that suggests that the cognitive profile of poor comprehenders is characterized by wider oral language skills. In particular, research suggests that poor comprehenders have impoverished vocabulary knowledge and experience difficulties acquiring new semantic representations. In addition, a deficit in grammar appears to be a stable feature of the poor comprehender profile that emerges before impairments in reading comprehension. Inferencing difficulties have been frequently observed in poor comprehender samples and research has investigated the extent to which these are linked to impairments in vocabulary, working memory and a drive for coherence. The pragmatic language skills of poor comprehenders are less well understood and there is a need to establish the extent to which poor comprehenders' performance on tests of reading comprehension and expressive language are linked to difficulties in understanding the social communication demands of the assessments.

Beyond language skills attention has been paid to aspects of executive functioning such as working memory and inhibition. Evidence suggests that working memory impairments in poor comprehenders may be specific to verbal working memory and a result of weaknesses in processing verbal information. Findings that poor comprehenders make more intrusion errors in verbal working memory tasks than controls, suggest that they also experience difficulties in inhibiting verbal information. Finally, metacognitive skills, such as the ability to recognize when comprehension has broken down and the extent to which an individual is concerned with whether a text makes sense, have been highlighted as areas of weakness in poor comprehenders.

There are therefore a number of possible cognitive factors that may be instrumental in causing the poor comprehender profile; these relate to the word level, the text base and the situation model of the CI model (Kintsch & Rawson, 2005). To further understand the relationships between these factors and their relative contributions to the reading comprehension difficulties of poor comprehenders, more longitudinal studies are needed and intervention studies designed to test specific casual hypotheses are likely to be vital in advancing understanding in this area.

IV. Biological Profile

Turning to the biological underpinnings of the poor comprehender profile, the following section presents two perspectives; the first focuses on genetic bases of reading comprehension and the second considers the neuroimaging literature.

A. GENETIC

It is now well established that reading skills are heritable and therefore that reading difficulties run in families. An accumulation of "family-risk" studies have suggested that between 30% and 50% of children whose parents have reading difficulties will also develop reading difficulties (Elbro, Borstrom, & Petersen, 1998; Gallagher, Frith, & Snowling, 2000;

Gilger, Pennington, & Defries, 1991; Scarborough, 1989, 1990, 1991). However, in addition to genes, families share environments so findings of familial aggregation of reading disorders alone are not sufficient to demonstrate heritability. Twin studies that compare the reading skills of monozygotic twins (who share 100% of their genes) with dyzygotic twins (who share approximately 50% of their genes) provide a sensitive design for unraveling genetic and environmental influences. These analyses assume that monozygotic and dyzygotic twins do not differ in the effects of shared environmental factors that contribute to their resemblance — the equal environments assumption (Plomin, DeFries, McClearn, & McGuffin, 2001).

Using the Colorado twin sample, Keenan, Betjemann, Wadsworth, DeFries, and Olson (2006) investigated the genetic and environmental etiology of individual differences in reading comprehension. They tested monozygotic and dyzygotic twins with multiple measures of word recognition, listening comprehension, and reading comprehension. Shared genetic influences on word recognition and reading comprehension were partly independent of shared genetic influences on listening and reading comprehension. The independence of these genetic influences on reading comprehension supports the simple view of reading, and the position that dyslexia and reading comprehension impairment are partly underpinned by different genetic etiologies. Quantitative genetic analyses such as this have contributed toward the view that reading is a continuous trait and that reading skills are significantly heritable across the distribution, as are the skills which underlie them.

Findings of family and twin studies showing the heritability of reading disorders pave the way for molecular genetic approaches to identify genes that may be associated with the poor comprehender profile. Although yet to be identified, it is likely that reading comprehension impairment, in common with other developmental cognitive disorders, is genetically heterogeneous and that the risk of inheriting this difficulty depends on the combined influence of many genes as well as environmental influences.

B. NEURAL

Little attention has been devoted to understanding the functional neuroanatomy of the poor comprehender profile and at present there are no clear neurological explanations. This is mainly because reading comprehension is a highly complex skill that relies on a large number of underlying skills and processes.

Some previous anatomical findings accord with the idea that comprehension and phonological deficits have different origins. For example, Leonard and colleagues found that brain measures in generally reading impaired students with poor reading comprehension more closely resembled those of children with specific language impairment than those of students with dyslexia (Leonard *et al.*, 2002). Leonard, Eckert, Given, Virginia, and Eden (2006) investigated neuroanatomical markers of reading difficulty in 22 children (aged 11–16 years) with reading and language impairments. They found that children with relatively smaller cerebral volume and auditory cortex and less marked leftward asymmetry of the planum temporal, plana, and cerebellar anterior lobe (negative risk indices) had weaker comprehension skills. Children with larger cerebral and auditory cortex and more marked asymmetries (positive risk indices) had poor word reading in the presence of relatively preserved comprehension. This study provides neuroanatomical support for the simple view of reading, and suggests that word reading and comprehension deficits are marked by neuroanatomical abnormalities on the opposite ends of the spectrum.

Most functional neuroimaging studies of sentence comprehension have been conducted with skilled readers (e.g., Buchweitz, Mason, Tomitch, & Just, 2009; Jobard, Vigneau, Mazoyer, & Tzourio-Mazaoyer, 2007). Findings from these studies have revealed that the comprehension of isolated words and sentences is associated with similar patterns of activation; however, the activation is more widespread for sentence comprehension which is associated with bilateral activation of the inferior frontal gyrus and the posterior superior and middle temporal gyri.

Specific literature in this area relating to individuals with reading difficulties is limited. Studies of individuals with decoding difficulties have reported abnormal patterns of activation during sentence comprehension (e.g., Helenius, Tarkiainen, Cornelissen, Hansen, & Salmelin, 1999; Meyler *et al.*, 2007). Particularly relevant is that these abnormalities are present in dyslexic groups not only when they are reading sentences but also when they are listening to sentences (Sabisch, Hahne, Glass, von Suchodoletz, & Friederici, 2006), that is, when reading accuracy or fluency is not constraining comprehension.

In a more recent fMRI study, Rimrodt *et al.* (2009) investigated the neurological underpinnings of reading and listening comprehension in children with dyslexia and age-matched peers. They reported that better word- and text-level reading fluency was associated with greater left occipitotemporal activation, near the area commonly referred to as the visual word form area. Impaired behavioral performance on oral language and reading comprehension abilities was associated with greater right hemisphere activation, particularly in the parietal lobe. More specifically, right supramarginal gyrus activation was negatively correlated with reading comprehension and receptive language, whereas right superior parietal lobe activation was negatively correlated with listening comprehension. This provides neurological support for the functional differences between reading and listening comprehension.

Taken together, findings from neuroanatomical and functional neuroimaging studies suggest that comprehension involves a highly distributed network of brain regions including aspects of the parietal lobes, the bilateral temporal poles and the right hemisphere. Future studies must investigate the neurological underpinnings of the poor comprehender profile so that a neurodevelopmental explanation for this reading profile can be reached. Such research must incorporate both behavioral and neurobiological findings for the purpose of developing a more complete understanding of the development of both normal and impaired reading comprehension.

V. Environmental Factors

Learning to read is strongly influenced by environmental experiences which interact with neurological development, including the home literacy environment, reading instruction, and print exposure (Cunningham & Stanovich, 1998; Phillips & Lonigan, 2009). A longitudinal study of the impact of home literacy environment on reading development was conducted by Senechal (2006). This study found that different styles of home literacy environment influence the development of different component reading skills; children who receive direct instruction about letters, sounds, and concepts of print tend to learn to decode more rapidly than children whose parents use reading as an opportunity for language interaction, but the latter group tends to fare better in the development of reading comprehension. In line with this, the prevalence of poor comprehenders tends to be significantly greater in more socially–economically deprived areas where the frequency of language interactions during reading and the quality and quantity of exposure to vocabulary may be lower (Hulme & Snowling, 2009).

It is also likely that the amount of reading practice a child undertakes affects the development of reading comprehension skills. Such that good readers are more likely to enjoy reading, and therefore practice more and gain further benefits. This is referred to as the "Matthew Effect" (Stanovich, 1986). For poor comprehenders, declining levels of motivation may feed into a downward spiral of poor reading comprehension and educational achievement. A decline in motivation may be particularly apparent at the shift from learning to read to reading to learn, since until this point they are as effective as peers at the task of decoding and will have received positive reinforcement from succeeding at the mechanics of reading. This illustrates a gene–environment interaction, whereby children who inherit a risk of reading difficulty place themselves in an environment that is not conducive to reading development.

At present, we are lacking systematic studies into the environmental factors that influence the development of the poor comprehender profile. Longitudinal studies are needed to assess the impact of the home and school literacy environments on the development of reading comprehension. Such research will be important in enabling us to provide parents with information about how best to promote reading comprehension development in the home.

VI. Interventions

There is a vast amount of literature on the topic of improving reading comprehension. Many of the studies available have a strong pedagogical focus and relate to techniques that can be applied to groups of typically developing children rather than to individuals who have specific difficulties in the area of reading comprehension. To provide some consensus, the National Reading Panel (2000) conducted a meta-analysis of the literature and identified specific approaches that they considered to have sufficient evidence bases to be regarded as effective. Of 203 studies reviewed, seven were deemed effective; comprehension monitoring, cooperative learning, graphic and semantic organizers (including story maps), story structure, question answering, question generation and summarization. The report also advocated the use of multiple strategy teaching in which teachers and children interact to apply a range of different strategies to the task of text comprehension. The results are echoed in a summary of intervention research by Duke, Pressley, and Hilden (2004) selected by Allington and McGill-Franzen (2009) it states:

"One of the most certain conclusions from the literature on comprehension strategies instruction is that long-term teaching of a small repertoire of strategies, beginning with teacher explanation and modeling with gradual release of responsibility to the student is very effective in promoting students' reading comprehension. This conclusion holds true for students with learning disabilities as well ..." (p. 512; Duke *et al.* 2004).

While this quote makes reference to students with learning disabilities it does not directly mention the needs of learners who have a poor comprehender profile. Although it is likely that poor comprehenders may benefit from the same approaches as other learners, support needs to be tailored to their particular strengths and weaknesses. It is likely that poor comprehenders will progress rapidly through reading schemes of books that place emphasis on successful word recognition and decoding and therefore they may be encountering texts that are far in advance of their comprehension ability. It is imperative that we understand the best ways to support poor comprehenders as they are at risk of becoming disillusioned with reading books that they do not understand.

Intervention studies have the potential to greatly advance our understanding of reading disorders; they are testing grounds for causal theories and can inform educational practice. The following section provides an overview of interventions designed to meet the needs of poor comprehenders and examines the contributions they have made to our understanding of the causes, characteristics, and remediation of the poor comprehender profile.

A. INFERENCE TRAINING

To determine the extent to which impairments in inferencing skills could be considered a causal factor in the poor comprehender profile, Yuill and Oakhill (1988) (replicated by McGee & Johnson, 2003), investigated the impact of inference awareness training on the reading comprehension skills of poor comprehenders and a comparison group of skilled comprehenders. Multiple training conditions were evaluated (inference awareness, standard comprehension, and rapid decoding). In the inference awareness training, children were required to describe the meaning of key words in relation to the context of the surrounding text. To supplement this, two additional strategies were taught: question generation and prediction. In the standard comprehension condition, participants answered questions and engaged in peer discussion, and in the rapid decoding condition, participants read words and passages at speed. Yuill and Oakhill (1988) demonstrated that poor comprehenders could make significant progress in reading comprehension as a result of relatively little additional support (6×45 min sessions). Following inference awareness training the poor comprehenders made an average improvement of 17 months on a standardized measure of reading comprehension (NARA, Neale, 1966). The children who received standard comprehension activities also made significant improvements in reading comprehension (on average 14 months); however, no significant improvements were shown by poor comprehenders who received rapid decoding. When the improvements made by the poor comprehenders were compared to those of the control group, only one significant difference was found. This was in the inference awareness training condition in which the poor comprehenders benefitted more than the skilled comprehenders. These findings have been taken to suggest that difficulties in making inferences may be a causal factor in the development of a poor comprehender profile. This study does not, however, rule out the possibility that other factors are of causal importance as gains in reading comprehension were also made by the poor comprehenders who received the standard comprehension activities.

Yuill and Joscelyne (1988) report two studies the second of which provides evidence that converges with the findings of Yuill and Oakhill (1988). Ten poor comprehenders and 10 skilled comprehenders were split into two conditions (training or no training). In a single session, the participants in the training condition were shown a short story that had been constructed specifically so that a high degree of inferencing was needed for understanding. They were then given a series of questions that required inferences to be made, and given some instructions on how to look for clues in the story. Once identified, the contributions of the clues to the answers were explained and children were then required to find and explain clues independently in relation to a new story. Following this session a series of eight experimental passages and questions were administered to examine the extent to which children had learnt and could apply the clue finding strategy. The authors reported a significant improvement in the ability to answer inference questions for the poor comprehender group relative to the untrained poor comprehender controls. The same comparison was not significant for the skilled comprehender group. At posttest there was also no significant difference in inferencing performance between the poor comprehender and skilled comprehender trained groups. This study shows that poor comprehenders can be taught to use clues in text to support inferencing skills.

B. MENTAL IMAGERY TRAINING

Historically, training to support the construction of mental images while reading has received quite a lot of research attention. Gambrell and Jaywitz (1993) provide an overview of this and present evidence to suggest that instructions to prompt mental image creation can improve story recall in typically developing fourth grade students. Oakhill and Patel (1991) evaluated the effectiveness of a mental imagery training program designed to improve the reading comprehension skills of poor comprehenders. They were influenced by Peters, Levin, McGivern, and Pressley (1985) who proposed that two forms of training may be beneficial. First, training to promote the creation of representational images (direct correspondence to the information present in the text) and, second, training to encourage the use of transformational images (specifically created to aid memory for the text). The program used by Oakhill and Patel consisted of three 20-min sessions delivered to small groups. Activities began with creating concrete representational drawings to accompany text (cartoon sequences and pictures to capture the main idea of the story); following this, children were encouraged to create and discuss similar pictures in their minds. Transformational style drawings were then introduced with the focus on depicting specific details to be remembered. In the final session of training no concrete drawings of either type were created, only mental images.

Twelve poor comprehenders and 12 skilled comprehenders (aged 9–10) were split between the training condition and a control condition (in which participants completed question answering activities). Using a bespoke measure that was based on the training material, it was found that poor comprehenders benefited more from imagery training than the control training, whereas skilled comprehenders' improvements did not differentiate according to training type. The authors suggest that "the ability to use imagery strategies may give poor comprehenders a way of helping to circumvent their memory limitations..." (p. 114) and that imagery training "... enables them, or forces them, to integrate information in the text in a way that they would not normally do" (p. 114). To test these hypotheses, it would be advantageous to examine the extent to which improvements as a result of mental imagery training are mediated by improvements in memory and integration skills. Oakhill and Patel (1991) acknowledge that further work is also necessary to establish which form of imagery training (representational or transformational) may be instrumental in the improvements in reading comprehension.

C. MULTIPLE STRATEGY BASED INTERVENTIONS

Many intervention approaches incorporate multiple strategies to reflect the multicomponential nature of reading comprehension. Two such approaches are the Visualizing/Verbalizing program (V/V; Bell, 1986) and Reciprocal Teaching (RT; Palincsar & Brown, 1984). The V/V program combines mental imagery activities with the creation of verbal summaries. In contrast, the RT program centers on peer and tutor discussion that encourages the use of four key strategies (Clarification, Summarization, Prediction, and Question Generation). Pearson (2009) argues

"Of all the approaches to strategy instruction that emerged in the 1980s and 1990s, none has had more direct impact than Reciprocal Teaching" (pg. 21). Its evidence base for effectiveness is strong (see Rosenshine & Meister, 1994 for review) and it includes two of the strategies highlighted in the National Reading Panel (2000) review.

Johnson-Glenberg (2000) investigated the extent to which these two programs improved the reading comprehension skills of poor comprehenders. Importantly, the design of this study incorporates an untreated control group and two training groups (V/V and RT). The training programs largely adhered to the procedures intended by Bell (1986) and Palincsar and Brown (1984). Training was given to small groups of children who received on average twenty-eight 30-min sessions within a 10-week period. Results revealed differential effects of the training programs on components of reading comprehension relative to controls. While the RT program led to gains in word recognition, question generation and factual understanding, the V/V program led to significant improvements in inference making. The finding that poor comprehenders who received the V/V training improved their inferencing skills warrants further investigation, as the nature of the link between inferencing and image making is not yet well understood.

Johnson-Glenberg (2005) extended this work by evaluating a web-based training program (3D-Readers), which was based around science texts and included visual (create a model) and verbal (question generation) strategies. Twenty poor comprehenders aged 11–13 years were selected by their teachers to take part. The study used a within subjects design (control condition followed by experimental condition). In the control condition (which took place over three sessions), participants read texts and unscrambled anagrams. In the experimental condition (which was conducted over five sessions), participants read texts which had activities based around the two strategies embedded within them. The results revealed significant improvements on bespoke measures of comprehension and reading strategy use as a result of the experimental condition. In addition, it was noted that poor comprehenders also changed in terms of rereading behavior, with more scroll backs happening during the experimental condition. Although this study is limited in sample size and the use of bespoke outcome measures, it is significant as it opens up a promising area for future development. It suggests that computer led activities can support the reading comprehension development of poor comprehenders. In line with this, Kamil and Chou (2009) conducted a brief review of the literature relating to the use of ICT in supporting reading comprehension and concluded that there is a small but reliable evidence base particularly from studies that have used computers to promote strategy use and metacognitive skills.

D. VOCABULARY TRAINING

If we accept that individuals who show the poor comprehender profile have weaknesses in the areas of vocabulary and semantics then the most logical step forward is to evaluate the extent to which interventions designed to promote vocabulary development and enrich semantic representations can alleviate difficulties in reading comprehension. There is a long history of research, which has explored the potential causal link between vocabulary and reading comprehension. A variety of hypotheses have been suggested to explain why vocabulary training might improve reading comprehension (Anderson & Freebody, 1981; Mezynski, 1983; Stahl & Nagy, 2006); these are outlined in a meta-analysis by Elleman, Lindo, Morphy, and Compton (2009) and a recent review by Baumann (2009).

In their meta-analysis, Elleman *et al.* (2009) reviewed 37 studies that explicitly sought to improve reading comprehension using vocabulary instruction. Of these, the studies which used a bespoke test of reading comprehension as their primary outcome measure demonstrated greater effects ($d=0.50$) than studies that used standardized tests ($d=0.10$). In general, effects were particularly strong for struggling readers. The authors had hoped that as a result of this analysis they would be in a position to make recommendations about the most effective methods of vocabulary instruction; however, they concluded that "... no matter what type of vocabulary instruction was used, it produced the same effect on comprehension as any other type of vocabulary instruction" (p. 35).

To meet the needs of diverse and heterogeneous populations of learners such as those with poor comprehender profiles, it might be most beneficial to implement a comprehensive vocabulary intervention which aims to approach the task of learning words from multiple angles. Graves (2006) has proposed a four-component framework designed to provide a comprehensive approach to vocabulary instruction. The components included are "(1) providing rich and varied language experiences; (2) teaching individual words; (3) teaching word learning strategies; and (4) fostering word consciousness" (p. 5). One approach that encompasses these components is the Multiple Context Learning (MCL) approach by Beck, McKeown, and Kucan (2002). This approach targets novel vocabulary items that are just above the age of acquisition expected for the child. These words are selected to be interesting, relevant and, most importantly, to enable the child to reflect on their own experiences of the word meaning. The approach fosters a prescriptive discussion of the new word in context between student(s) and tutor at which point a wide range of familiar and related vocabulary is activated and used in sentences. This creates an "umbrella" effect in which the novel difficult vocabulary item becomes associated with a series of well-known words which relate closely to the child's background knowledge. Using this method, the MCL approach aims to develop a rich representation of word meaning embedded in context.

To the best of our knowledge, there have been no studies that have specifically investigated the impact of vocabulary training on the reading comprehension ability of individuals who show a poor comprehender profile. However, we shall now provide details of our recent study that has used vocabulary training (based on the principles of Graves (2006) and MCL) as part of a multicomponential oral language (OL) program to facilitate reading comprehension in poor comprehenders.

E. EVALUATION OF INTERVENTIONS—A STUDY

Clarke, Snowling, Truelove, and Hulme (2010) is the first randomized controlled trial (RCT) to examine the effectiveness of different multicomponential packages in supporting the reading comprehension skills of poor comprehenders. The RCT design follows the CONSORT guidance (Moher, Schulz, & Altman, 2001), which is considered to be the gold standard in intervention research, as the random allocation of participants to groups ensures that additional influences are equally distributed across conditions. This reduces the possibility of results being confounded by unmeasured additional factors (e.g., school context, home environment). In Clarke *et al.*, we employed a waiting list control group whose progress was monitored alongside that of the participants receiving training; this group received support only after all posttest data collection was complete. This is an important feature that is missing in many intervention studies and allows authors to make claims about the extent to which training generates improvements above and beyond those that would normally be seen in individuals over that time period.

One hundred and sixty children aged 8–9 years with poor comprehender profiles were identified using a discrepancy criteria rather than a strict cutoff criteria. Eight children within each school were randomly allocated to one of three intervention groups or the control group. Three interventions were evaluated: (1) a text level program (TC), (2) an OL program, and (3) a combined program (COM). All three programs had

RT (Palincsar & Brown, 1984) at their core; in the case of the TC program this was applied to text, for the OL program it was applied to spoken language, and for the COM program it was applied to language in both domains. Similarly, narrative instruction (Beck & McKeown, 1981; Idol & Croll, 1987) was a component of all three programs, again the nature of this varied according to the domain of language that was being focused on. The OL program included two further components: vocabulary training and figurative language support. The vocabulary training closely followed the MCL approach of Beck *et al.* (2002) and incorporated additional features such as illustrations, verbal reasoning, graphic organizers/semantic maps (Nash & Snowling, 2006) and mnemonics (Levin, 1993). While the figurative language activities centered on developing participants' understanding of jokes (Yuill, 2007), riddles, metaphor, simile, and idiom. The TC program also included two additional components; inferencing from text and metacognitive strategies. The inferencing strategies were largely inspired by those used in Yuill and Oakhill (1988) and Yuill and Joscelyne (1988), but activities were extended to include many types of inference (lexical, bridging, elaborative, and evaluative). The metacognitive strategies taught were Reread (Garner, Chou Hare, Alexander, Haynes, & Winograd, 1984), Look-back (Garner, 1982), Think aloud (Farr & Conner, 2004), Mental imagery (Oakhill & Patel, 1991), and Explain and Reflect (McNamara, 2004). The COM program connected oral language and text-based activities in an integrated and naturalistic approach and included all components from the TC and OL programs.

The training consisted of pair (twice a week) and individual (once a week) sessions, each lasting approximately 30 minutes. These were led by trained teaching assistants, and total teaching time amounted to 30 hours of support over 20 weeks. The main outcome variable was reading comprehension, assessed using the NARA-II and the WIAT-II. All participants who received training performed significantly better than controls on the WIAT-II at posttest immediately following intervention. The NARA-II, however, failed to replicate this pattern of improvements, emphasizing the importance of using multiple measures of reading comprehension when attempting to capture gains resulting from intervention. Importantly, our study also included a further posttest, approximately 11 months after intervention had finished. Allington and McGill-Franzen (2009) highlight the issue of the lack of long-term evidence in reading comprehension intervention research and state "... there is less evidence that comprehension-focused interventions produce either autonomous use of comprehension strategies or longer-term improvements on comprehension proficiencies. The lack of evidence stems from the heavy reliance on smaller sample sizes and shorter-term intervention designs as well as limited attention to the "gold standard" of transfer of training to autonomous use ..." (p. 564)

At this posttest, we were able to demonstrate that initial gains made on the WIAT-II were lasting and a significant gain emerged for the OL group relative to the control group on the NARA-II. While improvements made by the TC and COM groups were sustained there was evidence that the OL group continued to make improvements on the WIAT-II, on average the OL group made approximately an eight standard score gain from pre-test to maintenance test.

To investigate the impact of vocabulary training on improvements in reading comprehension, we included measures of vocabulary at pretest and immediate posttest. The OL group made significant improvements relative to controls on a bespoke definition test of both taught and non-taught words, and a standardized measure of expressive vocabulary (Wechsler Abbreviated Scale of Intelligence, WASI, Wechsler, 1999). The COM group also made significant improvements relative to controls but only on the bespoke definition test of taught words. It is particularly noteworthy that the OL group significantly improved on the WASI vocabulary test, this was followed up at maintenance test but at this point it just failed to reach significance. Kamil and Chou (2009) recognize the difficulties with achieving gains on standardized tests of vocabulary and highlight that only two of the studies included in the National Reading Panel (2000) were successful in demonstrating such gains. We are there-fore encouraged by the present results and would suggest that they war-rant future attempts at replication.

To statistically account for the role of vocabulary gains in improvements in reading comprehension, a composite vocabulary measure (derived from both taught and non taught versions of the bespoke definition test) was entered as a potential mediator into a regression based structural equation model. The model revealed that vocabulary gains completely mediated improvements in reading comprehension in the COM group and partially accounted for reading comprehension gains in the OL group. This supports a causal relationship between oral language and reading compre-hension, and is consistent with the simple view (Gough & Tunmer, 1986); it suggests that to a certain extent the reading comprehension difficulties experienced by individuals who show the poor comprehender profile are due to weaknesses in vocabulary. However, it remains to be established precisely which aspects of vocabulary are crucial to this relationship.

We have demonstrated in Clarke *et al.* (2010) that the reading compre-hension difficulties of individuals with a poor comprehender profile can be significantly improved using multicomponential packages comprising activities that have an existing evidence base for effectiveness. It is of note that the intervention which generated the most significant and lasting gains in text comprehension was conducted in the spoken language domain and did not make explicit references to text. Importantly, the programs were delivered in school by teaching assistants demonstrating the pedagogical feasibility of these packages of intervention. A drawback of the multicomponential approach is that it becomes difficult to disentangle the different influences of specific aspects of the training programs. The extent to which precise conclusions can be drawn is constrained by the measures available to capture gains. In an ideal situation, there would be measures in place to assess all features and strategies in the programs; however, this is unfeasible, partly because of the time it would take to conduct such an assessment battery, and partly because some aspects are incredibly difficult to reliably and validly measure. This approach was taken to ensure interventions were educationally realistic and to maximize the chances of success for the participants involved and therefore the study fulfilled its aims.

VII. Summary and Future Directions

This chapter has considered the components of reading comprehension and how they can fractionate in reading disorders. The focus has been on providing a picture of current understanding of the poor comprehender profile as well as introducing the issues and key questions pertinent to this field. It is recognized that the simple view (Gough & Tunmer, 1986) is useful in distinguishing the dyslexic and poor comprehender profiles and in capturing the basic skills fundamental to successful reading. The simple view, however, fails to explain the complexities in reading comprehension, and to do this, alternative models such as the CI model (Kintsch & Rawson, 2005) need to be considered.

To further our understanding of how the poor comprehender profile emerges and how different factors interact over time, a developmental theory is needed to account for the levels of explanation outlined in Morton and Frith (1995). To achieve this, additional research into the biological, environmental, and social factors that contribute to the poor comprehender profile is necessary. Moreover, we have highlighted a number of issues surrounding the assessment and identification of poor comprehenders; it is imperative that these issues are reconciled in order for this field of research to advance.

Paula J. Clarke et al.

In our view, the ultimate goal of reaching a full understanding of the causes of the poor comprehender profile is to inform theory-driven intervention approaches. To date, there has been surprisingly little research attention focusing on methods for supporting children with poor comprehender profiles. On the basis of the evidence so far, two strategies seem particularly important in this population; early studies by Oakhill and colleagues established the effectiveness of inference awareness training and our recent study has underscored the importance of vocabulary training. This recognizes the assertion by Cain *et al.* (2004) that "... there is a need to study the relation between learning from context, comprehension skill and vocabulary acquisition over time" (p. 680). Future intervention approaches could further examine the relationship between inferencing and vocabulary acquisition and explore the feasibility of integrating the methods that give focus to understanding using context. Given poor comprehenders' weaknesses in listening comprehension, an extension of this would be to investigate the training of inferencing in the spoken language domain.

It is important to recognize that children can score poorly on a test of reading comprehension for a multitude of reasons. Such reasons may be features of specific clinical profiles. We have highlighted that it may be useful for future research to explore the relationships between inattention and reading comprehension in individuals with ADHD. A further avenue for investigation is the incidence of poor comprehender profiles in individuals with Autism Spectrum Disorder (ASD).

There is broad agreement that children with ASD have impaired reading and listening comprehension relative to word reading ability (Nation, Clarke, Wright, & Williams, 2006). However, the causes of this profile are poorly understood. One hypothesis is that impairments in semantic skills directly compromise the development of reading comprehension, similar to poor comprehenders who are otherwise developing normally (McCleery *et al.*, 2010; Luyster, Kadlec, Carter, & Tager-Flusberg, 2008). However, impairments in the ability to attribute mental states to others (ToM), engaging in joint attention, and pragmatic language skill are also well documented in ASD (Adams, Green, Gilchrist, & Cox, 2002; Baron-Cohen, Leslie, & Frith, 1985; Frith, 2001; Tager-Flusberg, 2007). Such impairments may lead to children engaging less well with texts and being less concerned with maintaining a high standard of coherence. Children with ASD may lack a desire to understand authors' or speakers' intent, which may decrease their motivation to process language for meaning. Beyond semantic and socio-communicative language impairments, superior attention to detail may also play a role in the poor comprehender profile observed in ASD (Burack, Enns, Stauder, Mottron,

& Randolph, 1997; Joseph, Keehn, Connolly, Wolfe, & Horowitz, 2009; Kemner, Lizet, Herman, & Ignace, 2008; O'Riordan, Plaisted, Driver, & Baron-Cohen, 2001). Enhanced visual discrimination may have advantages for the early stages of decoding development (e.g., when acquiring letter-knowledge) but this same superiority may divert attention away from the communication of meaning. It should also be noted that impairments in executive skills such as inhibition, planning, and monitoring (Hughes, 1996; Ozonoff & Miller, 1996) may also contribute toward the poor comprehender profile in ASD. Although speculative at present, it appears that there may be a number of risk factors for comprehension difficulties that are deserving of further research. Studies that identify predictors of growth in decoding and comprehension in other disorders characterized by comprehension impairments such as ASD will lead to important insights into the routes to the poor comprehender profile.

Reading is a unique human accomplishment that is critical for educational achievement, mental capital, and well being for all children. Although the purpose of the reading process is to extract meaning, children who demonstrate specific weaknesses in understanding what they read often go unnoticed in the classroom because they can read aloud with ease and accuracy. Increasing awareness of the reading profile of poor comprehenders and making efforts to improve the identification and support provision for these children may be instrumental in raising attainment across the curriculum.

REFERENCES

Adams, C., Green, J., Gilchrist, A., & Cox, A. (2002). Conversational behavior of children with Asperger syndrome and conduct disorder. *Journal of Child Psychology and Psychiatry, 43*, 679–690.

Allington, R. L., & McGill-Franzen, A. (2009). Comprehension difficulties among struggling readers. In S. E. Israel & G. G. Duffy (Eds.), *Handbook of research on reading comprehension* (pp. 551–568). NY: Taylor & Francis.

Anderson, R. C., & Freebody, P. (1981). Vocabulary knowledge. In J. T. Guthrie (Ed.), *Comprehension and teaching: Research reviews* (pp. 77–117). Newark, DE: International Reading Association.

Baddeley, A. D. (1992). Consciousness and working memory. *Consciousness and Cognition, 1*, 3–6.

Baddeley, A. D. (2007). *Working memory, thought and action.* Oxford: Oxford University Press.

Barnes, M. A., Faulkner, H., Wilkinson, M., & Dennis, M. (2004). Meaning construction and integration in children with hydrocephalus. *Brain and Language, 89*, 47–56.

Baron-Cohen, S., Leslie, A. M., & Frith, U. (1985). Does the autistic child have a theory of mind? *Cognition, 21*, 37–46.

Baumann, J. F. (2009). Vocabulary and reading comprehension: The nexus of meaning. In S. E. Israel & G. G. Duffy (Eds.), *Handbook of research on reading comprehension* (pp. 323–346). NY: Taylor & Francis.

Beck, I. L., & McKeown, M. G. (1981). Developing questions that promote comprehension: The story map. *Language Arts, 58*(8), 913–918.

Beck, I. L., McKeown, M. G., & Kucan, L. (2002). *Bringing words to life: Robust vocabulary instruction.* New York: Guilford Press.

Bell, N. (1986). *Visualizing and verbalizing for language comprehension and thinking.* Paso Robles, CA: Academy of Reading Publications.

Bishop, D. V. M. (1982). *Test for reception of grammar. Medical research council.* Oxford UK: Chapel Press.

Bishop, D. V. M. (1998). Development of the children's communication checklist (CCC): A method for assessing qualitative aspects of communicative impairment in children. *Journal of Child Psychology and Psychiatry, 39*(6), 879–891.

Bishop, D. V. M. (2003). *The test for reception of grammar, version 2 (TROG-2).* London: Psychological Corporation.

Bowyer-Crane, C., & Snowling, M. J. (2005). Assessing children's inference generation: What do tests of reading comprehension measure? *The British Journal of Educational Psychology, 75,* 189–201.

Bowyer-Crane, C., & Snowling, M. J. (2008). Turning frogs into princes: Can children make inferences from fairy tales? *Reading and Writing, 23,* 19–29.

Brock, S. E., & Knapp, P. K. (1996). Reading comprehension abilities of children with attention-deficit/hyperactivity disorder. *Journal of Attention Disorders, 1,* 178–184.

Buchweitz, A., Mason, R. A., Tomitch, L. M. B., & Just, M. A. (2009). Brain activation for reading and listening comprehension: An fMRI study of modality effects and individual differences in language comprehension. *Psychology & Neuroscience, 2,* 111–123.

Burack, J. A., Enns, J. T., Stauder, J. E. A., Mottron, L., & Randolph, B. (1997). Attention and autism: Behavioral and electrophysiological evidence. In D. J. Cohen & F. R. Volkmar (Eds.), *Handbook of autism and pervasive developmental disorders* (pp. 226–247). New York: Wiley.

Cain, K. (2006). Individual differences in children's memory and reading comprehension: An investigation of semantic and inhibitory deficits. *Memory, 14,* 553–569.

Cain, K., & Oakhill, J. V. (1999). Inference making ability and its relation to comprehension failure. *Reading and Writing, 11,* 489–503.

Cain, K., & Oakhill, J. V. (2006). Profiles of children with specific reading comprehension difficulties. *The British Journal of Educational Psychology, 76,* 683–696.

Cain, K., & Oakhill, J. (2009). Reading comprehension development from 8–14 years: The contribution of component skills and processes. In R. K. Wagner, C. Schatschneider & C. Phythian-Sence (Eds.), *Beyond decoding: The behavioral and biological foundations of reading comprehension.* NY: The Guildford Press.

Cain, K., Oakhill, J. V., Barnes, M. A., & Bryant, P. E. (2001). Comprehension skill, inference making ability, and their relation to knowledge. *Memory & Cognition, 29,* 850–859.

Cain, K., Oakhill, J. V., & Lemmon, K. (2004). Individual differences in the inference of word meanings from context: The influence of reading comprehension, vocabulary knowledge, and memory capacity. *Journal of Educational Psychology, 96,* 671.

Cain, K., Oakhill, J. V., & Lemmon, K. (2005). The relation between children's reading comprehension level and their comprehension of idioms. *Journal of Experimental Child Psychology, 90,* 65–87.

Catts, H. W., Adlof, S. M., & Weismer, S. E. (2006). Language deficits in poor comprehenders: A case for the simple view of reading. *Journal of Speech, Language, and Hearing Research, 49,* 278–293.

Catts, H. W., Hogan, T. P., & Fey, M. E. (2003). Subgrouping poor readers on the basis of individual differences in reading-related abilities. *Journal of Learning Disabilities, 36*(2), 151.

Clarke, P. J., Snowling, M. J., Truelove, E., & Hulme, C. (2010). Ameliorating children's reading comprehension difficulties: A randomised controlled trial. *Psychological Science, 21*, 1106–1116.

Cragg, L., & Nation, K. (2006). Exploring written narrative in children with poor reading comprehension. *Educational Psychology, 21*, 55–72.

Cunningham, A. E., & Stanovich, K. E. (1998). What reading does for the mind. *American Educator, 22*, 8–15.

Daneman, M., & Carpenter, P. A. (1980). Individual differences in working memory and reading. *Journal of Verbal Learning and Verbal Behavior, 19*(4), 450–466.

De Beni, R., & Palladino, P. (2000). Intrusion errors in working memory tasks. Are they related to reading comprehension ability? *Learning and Individual Differences, 12*, 131–143.

De Beni, R., Palladino, P., Pazzaglia, F., & Cornoldi, C. (1998). Increases in intrusion errors and working memory deficit in poor comprehenders. *The Quarterly Journal of Experimental Psychology, 51a*, 305–320.

Duke, N. K., Pressley, M., & Hilden, K. (2004). Difficulties with reading comprehension. In C. A. Stone, E. R. Silliman, B. J. Ehren & K. Apel (Eds.), *Handbook of language and literacy: Development and disorders* (pp. 501–520). New York: Guilford.

Elbro, C., Borstrom, I., & Petersen, D. K. (1998). Predicting dyslexia from kindergarten: The importance of distinctness of phonological representations of lexical items. *Reading Research Quarterly, 33*, 36–60.

Elleman, A. M., Lindo, E. J., Morphy, P., & Compton, D. L. (2009). The impact of vocabulary instruction on passage-level comprehension of school-age children: A meta-analysis. *Journal of Research on Educational Effectiveness, 2*(1), 1–44.

ESRC Seminar series 'Reading comprehension: From theory to practice' report (unpublished) reading comprehension: Nature, assessment and teaching. Retrieved August 1, 2010 from http://www.york.ac.uk/res/crl/downloads/ESRCcomprehensionbooklet.pdf.

Farr, R., & Conner, J. (2004). *Using think-alouds to improve reading comprehension. http://www.readingrockets.org/article/102 Accessed 01.07.10.*

Frith, U. (2001). Mind blindness and the brain in autism. *Neuron, 32*, 969–979.

Gallagher, A. M., Frith, U., & Snowling, M. J. (2000). Precursors of literacy delay among children at genetic risk of dyslexia. *Journal of Child Psychology and Psychiatry, 41*, 202–213.

Gambrell, L. N., & Jaywitz, P. B. (1993). Mental imagery, text illustrations, and children's comprehension and recall. *Reading Research Quarterly, 28*, 264–273.

Garner, R. (1982). Resolving comprehension failure through text look backs: Direct training and practice effects among good and poor comprehenders in grades six and seven. *Reading Psychology, 3*(3), 221–223.

Garner, R., Chou Hare, V., Alexander, P., Haynes, J., & Winograd, P. (1984). Inducing use of a text lookback strategy among unsuccessful readers. *American Educational Research Journal, 21*(4), 789–798.

Gernsbacher, M. A. (1993). Less skilled readers have less efficient suppression mechanisms. *Psychological Science, 4*, 294–298.

Gernsbacher, M. A., & Faust, M. E. (1991). The mechanism of suppression: A component of general comprehension skill. *Journal of Experimental Psychology. Learning, Memory, and Cognition, 17*, 245–262.

Gernsbacher, M. A., Varner, K. R., & Faust, M. (1990). Investigating differences in general comprehension skill. *Journal of Experimental Psychology: Learning, Memory, and Cognition, 16,* 430–445.

Gilger, J. W., Pennington, B. F., & Defries, J. C. (1991). Risk for reading disability as a function of parental history in three family studies. *Reading and Writing: An Interdisciplinary Journal, 3,* 299–313.

Gough, P. B., & Tunmer, W. E. (1986). Decoding, reading, and reading disability. *Remedial and Special Education, 7,* 6–10.

Graves, M. F. (2006). *The vocabulary book: Learning & instruction.* New York: Teacher's College Press.

Guthrie, J. T., & Wigfield, A. (2000). Engagement and motivation in reading. In M. L. Kamil, P. B. Mosenthal, D. P. Pearson & R. Barr (Eds.), *Handbook of reading research* (Vol. 3). Mahwah, NJ: Lawrence Erlbaum Associates.

Guthrie, J. T., Wigfield, A., Metsala, J., & Cox, K. E. (1999). Predicting text comprehension and reading activity with motivational and cognitive variables. *Scientific Studies of Reading, 3,* 231–256.

Hasher, L., & Zacks, R. T. (1979). Automatic and effortful processes in memory. *Journal of Experimental Psychology, 108,* 356–388.

Helenius, P., Tarkiainen, A., Cornelissen, P., Hansen, P. C., & Salmelin, R. (1999). *Dissociation of normal feature analysis and deficient processing of letter-strings in dyslexic adults. Cerebral Cortex 9,* 476–483.

Hoffman, J. V. (2009). In search of the "simple view" of reading. In S. E. Israel & G. G. Duffy (Eds.), *Handbook of research on reading comprehension* (pp. 54–65). NY: Taylor & Francis.

Hoover, W. A., & Gough, P. B. (1990). The simple view of reading. *Reading and Writing: An Interdisciplinary Journal, 2,* 127–160.

Hughes, C. (1996). Control of action and thought: Normal development and dysfunction in autism. *Journal of Child Psychology and Psychiatry, 37,* 229–236.

Hulme, C., & Snowling, M. J. (2009). *Developmental disorders of language, learning and cognition.* UK: Wiley-Blackwell.

Idol, L., & Croll, V. J. (1987). Story-mapping training as a means of improving reading comprehension. *Learning Disability Quarterly, 10,* 214–229.

Jobard, G., Vigneau, M., Mazoyer, B., & Tzourio-Mazaoyer, N. (2007). Impact of modality and linguistic complexity during reading listening tasks. *NeuroImage, 34,* 784–800.

Johnson-Glenberg, M. C. (2000). Training reading comprehension in adequate decoders/poor comprehenders: Verbal vs. visual strategies. *Journal of Educational Psychology, 92,* 772–782.

Johnson-Glenberg, M. C. (2005). Web-based training of metacognitive strategies for text comprehension: Focus on poor comprehenders. *Reading and Writing, 18,* 755–786.

Joseph, R. M., Keehn, B., Connolly, C., Wolfe, J. M., & Horowitz, T. S. (2009). Why is visual search superior in autism spectrum disorder? *Developmental Science, 12,* 1083–1096.

Joshi, R. Malatesha, & Aaron, (2000). The component model of reading: Simple view of reading made a little more complex. *Reading Psychology, 21,* 85–97.

Kamil, M. L., & Chou, H. K. (2009). Comprehension and computer technology: Past results, current knowledge, and future promises. In S. E. Israel & G. G. Duffy (Eds.), *Handbook of research on reading comprehension* (pp. 289–304). NY: Taylor & Francis.

Keenan, J. M., & Betjemann, R. S. (2006). Comprehending the Gray Oral Reading Test without reading it: Why comprehension tests should not include passage-independent items. *Scientific Studies of Reading, 10,* 363–380.

Keenan, J., Betjemann, R., & Olson, R. (2008). Reading comprehension tests vary in the skills they assess: Differential dependence on decoding and oral comprehension. *Scientific Studies of Reading, 12,* 281–300.

Keenan, J. M., Betjemann, R. S., Wadsworth, S. J., DeFries, J. C., & Olson, R. K. (2006). Genetic and environmental influences on reading and listening comprehension. *Journal of Research in Reading, 29,* 75–91.

Kemner, C., Lizet, E., Herman, E., & Ignace, H. (2008). Brief report: Eye movements during visual search tasks indicate enhanced stimulus discriminability in subjects with PDD. *Journal of Autism and Developmental Disorders, 38,* 553–557.

Kintsch, W., & Rawson, K. A. (2005). Comprehension. In M. J. Snowling & C. Hulme (Eds.), *The science of reading: A handbook* (pp. 209–226). Malden, MA: Blackwell.

Landi, N., & Perfetti, C. A. (2007). An electrophysiological investigation of semantic and phonological processing in skilled and less skilled comprehenders. *Brain and Language, 102,* 30–45.

Leather, C. V., & Henry, L. A. (1994). Working memory span and phonological awareness tasks as predictors of early reading ability. *Journal of Experimental Child Psychology, 58,* 88–111.

Leonard, C., Eckert, M., Given, B., Virginia, B., & Eden, G. (2006). Individual differences in anatomy predict reading and oral language impairments in children. *Brain, 129,* 3329–3342.

Leonard, L. B., Miller, C. A., Deevy, P., Rauf, L., Gerber, E., & Charest, M. (2002). Production operations and the use of nonfinite verbs by children with specific language impairment. *Journal of Speech, Language, and Hearing Research, 45,* 744–758.

Levin, J. R. (1993). Mnemonic strategies and classroom learning: A twenty-year report card. *Elementary School Journal, 94,* 235.

Lindsay, G., Dockrell, J. E., & Mackie, C. (2008). Vulnerability to bullying in children with a history of specific speech and language difficulties. *European Journal of Special Needs Education, 23,* 1–16.

Luyster, R., Kadlec, M., Carter, A., & Tager-Flusberg, H. (2008). Language assessment and development in toddlers with autism spectrum disorders. *Journal of Autism and Developmental Disorders, 38,* 1426–1438.

McCleery, J. P., Ceponiene, R., Burner, K. M., Townsend, J., Kinnear, M., & Schreibman, L. (2010). Neural correlates of verbal and non-verbal semantic integration in children with autism spectrum disorders. *Journal of Child Psychology Psychiatry and Allied Disciplines, 51,* 277–286.

McGee, A., & Johnson, H. (2003). The effect of inference training on skilled and less skilled comprehenders. *Educational Psychology, 23,* 49–59.

McNamara, D. S. (2004). SERT: Self-explanation reading training. *Discourse Processes, 38,* 1–30.

McNamara, T. P. (2005). *Semantic priming: Perspectives from memory and word recognition.* New York: Psychology Press.

Meyer, D. E., & Schvaneveldt, R. W. (1971). Facilitation in recognizing pairs of words: Evidence of a dependence between retrieval operations. *Journal of Experimental Psychology, 90,* 227–234.

Meyler, A., Keller, T. A., Cherkassky, V. L., Lee, D., Hoeft, F., Whitfield-Gabrieli, S., *et al.* (2007). Brain activation during sentence comprehension among good and poor readers. *Cerebral Cortex, 17,* 2780–2787.

Mezynski, K. (1983). Issues concerning the acquisition of knowledge: Effects of vocabulary training on reading comprehension. *Review of Educational Research, 53,* 253–273.

Moher, D., Schulz, K. F., & Altman, D. G. (2001). The CONSORT statement: Revised recommendations for improving the quality of reports of parallel-group randomised trials. *Lancet, 357,* 1191–1194.

Morton, J., & Frith, U. (1995). Causal modelling a structural approach to developmental psychopathology. In D. Cicchetti & D. J. Cohen (Eds.), *Developmental psychopatholgy.* New York: Wiley.

Nagy, W., & Scott, J. (2000). Vocabulary Processes. In M. Kamil, P. Mosenthal, P. D. Pearson & R. Barr (Eds.), *Handbook of reading research* (Vol. 3, pp. 269–284). Mahwah, NJ: Erlbaum.

Nash, H., & Snowling, M. J. (2006). Teaching new words to children with poor existing vocabulary knowledge: A controlled evaluation of the definition and context methods. *International Journal of Language & Communication Disorders, 41*, 335–354.

Nation, K. (1999). Reading skills in hyperlexia: A developmental perspective. *Psychological Bulletin, 125*, 338.

Nation, K. (2009). Reading comprehension and vocabulary: what's the connection? In R. K. Wagner, C. Schatschneider & C. Phythian-Sence (Eds.), *Beyond decoding: The behavioral and biological foundations of reading comprehension.* NY: The Guildford Press.

Nation, K., Adams, J. W., Bowyer-Crane, C. A., & Snowling, M. J. (1999). Working memory deficits in poor comprehenders reflect underlying language impairments. *Journal of Experimental Child Psychology, 73*, 139–158.

Nation, K., Clarke, P., Marshall, C. M., & Durand, M. (2004). Hidden language impairments in children: Parallels between poor reading comprehension and specific language impairment? *Journal of Speech, Language, and Hearing Research, 47*, 199.

Nation, K., Clarke, P., Wright, B., & Williams, C. (2006). Patterns of reading ability in children with autism spectrum disorder. *Journal of Autism and Developmental Disorders, 36*, 911–919.

Nation, K., Cocksey, J., Taylor, J. S. H., & Bishop, V. M. (2010). A longitudinal investigation of early reading and language skills in children with poor reading comprehension. *Journal of Child Psychology and Psychiatry,* (online early view).

Nation, K., Marshall, C., & Altmann, G. (2003). Investigating individual differences in children's real-time sentence comprehension using language-mediated eye movements. *Journal of Experimental Child Psychology, 86*, 314–329.

Nation, K., Marshall, C., & Snowling, M. J. (2001). Phonological and semantic contributions to children's picture naming skill: Evidence from children with developmental reading disorders. *Language and Cognitive Processes, 16*, 241–259.

Nation, K., & Snowling, M. J. (1997). Assessing reading difficulties: The validity and utility of current measures of reading skill. *The British Journal of Educational Psychology, 67*, 359–370.

Nation, K., & Snowling, M. J. (1998a). Semantic processing and the development of word recognition skills: Evidence from children with reading comprehension difficulties. *Journal of Memory and Language, 39*, 85–101.

Nation, K., & Snowling, M. J. (1998b). Individual differences in contextual facilitation: Evidence from dyslexia and poor reading comprehension. *Child Development, 69*, 996–1011.

Nation, K., & Snowling, M. J. (1999). Developmental differences in sensitivity to semantic relations among good and poor comprehenders: Evidence from semantic priming. *Cognition, 70*, B1–B13.

Nation, K., Snowling, M. J., & Clarke, P. J. (2007). Dissecting the relationship between language skills and learning to read: Semantic and phonological contributions to new vocabulary learning in children with poor reading comprehension. *Advances in Speech-Language Pathology, 9*, 131–139.

National Reading Panel (2000). *Teaching children to read: An evidence based assessment of the scientific research literature on reading and its implications for reading instruction.* National Institute of Child Health and Human Development. http://www.nationalreadingpanel.org/default.htm.

Neale, M. D. (1966). *The Neale analysis of reading ability.* London: Macmillan.

Neale, M. D. (1997). *Neale analysis of reading ability – Second revised British edition (NARA II).* London: NFER Nelson.

Neely, J. H. (1991). Semantic priming effects in visual word recognition: A selective review of current findings and theories. In D. Besner & G. W. Humphreys (Eds.), *Basic processes in reading, visual word recognition* (pp. 264–336). New Jersey: Lawrence Erlbaum Associates.

Nippold, M. A. (2002). Lexical learning in school-age children, adolescents, and adults: A process where language and literacy converge. *Journal of Child Language, 29*, 474–478.

O'Riordan, M. A., Plaisted, K. C., Driver, J., & Baron-Cohen, S. (2001). Superior visual search in autism. *Journal of Experimental Psychology: Human Perception and Performance, 27*, 719–730.

Oakhill, J. V. (1981). *Children's reading comprehension.* University of Sussex.

Oakhill, J. V. (1982). Constructive processes in skilled and less skilled comprehenders' memory for sentences. *British Journal of Psychology, 73*, 13–20.

Oakhill, J. V. (1983). Reading comprehension skill and detection errors on the letter *t. First Language, 3*, 111–120.

Oakhill, J. V., Hartt, J. & Samols, D. (1996). *Comprehension monitoring and working memory in good and poor comprehenders.* Paper presented at the 14th Biennial ISSBD Conference, Quebec City.

Oakhill, J. V., & Cain, K. (2007). Issues of causality in children's reading comprehension. In D. S. McNamara (Ed.), *Reading comprehension strategies: Theories, interventions and technologies.* Mahwah, NJ: Lawrence Erlbaum Associates, Inc.

Oakhill, J. V., Cain, K., & Bryant, P. E. (2003). The dissociation of word reading and text comprehension: Evidence from component skills. *Language and Cognitive Processes, 18*, 443–468.

Oakhill, J. V., & Patel, S. (1991). Can imagery training help children who have comprehension problems? *Journal of Research in Reading, 14*, 106–115.

Orton Dyslexia Society. (1994). cited in M. Snowling (2000) *Dyslexia* (pp. 24). Malden, MA: Blackwell Publishers.

Osterhout, L. (2000). On space, time and language: For the next century, timing is (almost) everything. *Brain and Language, 71*, 175–177.

Ozonoff, S., & Miller, J. N. (1996). An exploration of right-hemisphere contributions to the pragmatic impairments of autism. *Brain and Language, 52*, 411–434.

Palincsar, A. S., & Brown, A. L. (1984). Reciprocal teaching of comprehension-fostering and comprehension-monitoring activities. *Cognition and Instruction, 1*, 117–175.

Palladino, P., Cornoldi, C., De Beni, R., & Pazzaglia, F. (2001). Working memory and updating processes in reading comprehension. *Memory & Cognition, 29*, 344.

Pearson, P. D. (2009). The roots of reading comprehension instruction. In S. E. Israel & G. G. Duffy (Eds.), *Handbook of research on reading comprehension* (pp. 3–31). NY: Taylor & Francis.

Perfetti, C. (2007). Reading ability: Lexical quality to comprehension. *Scientific Studies of Reading, 11*, 357–383.

Perfetti, C. A., Landi, N., & Oakhill, J. V. (2005). The acquisition of reading comprehension skill. In M. J. Snowling & C. Hulme (Eds.), *The science of reading: A handbook* (pp. 227–247). Oxford, England: Blackwell.

Perfetti, C. A., Marron, M. A., & Foltz, P. W. (1996). Sources of comprehension failure: Theoretical perspectives and case studies. In C. Cornoldi & J. Oakhill (Eds.), *Reading comprehension difficulties: Processes and intervention* (pp. 137–166). Mahwah, NJ: Lawrence Erlbaum Associates, Inc.

Peters, E. E., Levin, J. R., McGivern, J. E., & Pressley, M. (1985). Further comparison of representational and transformational prose-learning imagery. *Journal of Educational Psychology, 77*, 129–136.

Phillips, B., & Lonigan, C. (2009). Variations in the home literacy environment of preschool children: A cluster analytic approach. *Scientific Studies of Reading, 13*, 146–174.

Pimperton, H., & Nation, K. (2010). Suppressing irrelevant information from working memory: Evidence for domain-specific deficits in poor comprehenders. *Journal of Memory and Language, 62*, 380–391.

Plomin, R., DeFries, J. C., McClearn, G. E., & McGuffin, P. (2001). *Behavioral genetics* (4th ed.). New York: Worth.

Pugzles-Lorch, E., Milich, R., & Sanchez, R. P. (1998). Story comprehension in children with ADHD. *Clinical Child and Family Psychology Review, 1*, 163–178.

Ricketts, J., Bishop, D. V. M., & Nation, K. (2008). Investigating orthographic and semantic aspects of word learning in poor comprehenders. *Journal of Research in Reading, 31*, 117–135.

Rimrodt, S. L., Clements, A. M., Pugh, K. R., Courteney, S. M., Gaur, P., Pekar, J. J., et al. (2009). *Functional MRI of sentence comprehension in children with dyslexia: Beyond word recognition.* Oxford University Press.

Rose, J. (2009). *Identifying and teaching children and young people with dyslexia and literacy difficulties.* Department for Children, Schools and Families. Retrieved from *http:// publications.education.gov.uk/default.aspx?PageFunction=productdetails&PagMode= publications&ProductId=DCSF-00659-2009.*

Rosenshine, B., & Meister, C. (1994). Reciprocal teaching: A review of the research. *Review of Educational Research, 64*, 479–530.

Sabisch, B., Hahne, A., Glass, E., von Suchodoletz, W., & Friederici, A. D. (2006). Auditory language comprehension in children with developmental dyslexia: Evidence from event-related brain potentials. *Journal of Cognitive Neuroscience, 18*, 1676–1695.

Scarborough, J. S. (1989). Prediction of reading disability from familial and individual differences. *Journal of Educational Psychology, 81*, 101–108.

Scarborough, H. S. (1990). Very early language deficits in dyslexic children. *Child Development, 61*, 1728–1743.

Scarborough, H. S. (1991). Early syntactic development of dyslexic children. *Annals of Dyslexia, 41*, 207–220.

Seigneuric, A., & Ehrlich, M.-F. (2005). Contribution of working memory capacity to children's reading comprehension: A longitudinal investigation. *Reading and Writing, 18*, 617–656.

Senechal, M. (2006). Testing the home literacy model: Parent involvement in kindergarten is differentially related to grade 4 reading comprehension, fluency, spelling, and reading for pleasure. *Scientific Studies of Reading., 10*, 59–87.

Shefelbine, J. L. (1990). Student factors related to variability in learning word meanings from context. *Journal of Reading Behavior, 22*, 71–97.

Snowling, M. J. (2000). *Dyslexia.* Malden, MA: Blackwell Publishers.

Snowling, M. J., Stothard, S. E., Clarke, P., Bowyer-Crane, C., Harrington, A., Truelove, E., et al. (2009). *York assessment of reading for comprehension: Passage reading.* London: GL Assessment.

Snowling, M. J., Stothard, S. E., Clarke, P., Bowyer-Crane, C., Harrington, A., Truelove, E., et al. (2010). *York assessment of reading for comprehension: Passage reading—Secondary version.* London: GL Assessment.

Spooner, A. L., Baddeley, A. D., & Gathercole, S. E. (2004). Can reading accuracy and comprehension be separated in the Neale analysis of reading ability? *The British Journal of Educational Psychology, 74*, 187–204.

Stahl, S. A., & Nagy, W. E. (2006). *Teaching word meanings*. Mahwah, N.J: Lawrence Erlbaum.

Stanovich, Keith E. (1986). Matthew effects in reading: Some consequences of individual differences in the acquisition of literacy. *Reading Research Quarterly, 21*, 360–407.

Sternberg, R. J., & Powell, J. S. (1983). Comprehending verbal comprehension. *The American Psychologist, 38*, 878–893.

Stothard, S. E., & Hulme, C. (1992). Reading comprehension difficulties in children: The role of language comprehension and working memory skills. *Reading and Writing, 4*, 245–256.

Stothard, S. E., & Hulme, C. (1995). A comparison of phonological skills in children with reading comprehension difficulties and children with decoding difficulties. *Journal of Child Psychology and Psychiatry and Allied Disciplines, 36*, 399–408.

Swinney, D. A. (1979). Lexical access during sentence comprehension: (Re)consideration of context effects. *Journal of Verbal Learning and Verbal Behaviour, 18*, 645–659.

Swinney, D., & Prather, P. (1989). On the comprehension of lexical ambiguity by young children: Investigations into the development of mental modularity. In D. Gorfein (Ed.), *Resolving semantic ambiguity*. Springer-Verlag: New York.

Taft, M. (1991). *Reading and the mental lexicon*. East Sussex, UK: Lawrence Erlbaum Associates.

Tager-Flusberg, H. (2007). Evaluating the theory of mind hypothesis of autism. *Current Directions in Psychological Science, 16*, 311–315.

van den Broek, P. W., Risden, K., & Husebye-Hartman, E. (1995). The role of readers standards for coherence in the generation of inferences during reading. In R. F. Lorch & E. J. O'Brien (Eds.), *Sources of coherence in reading* (pp. 353–373). Mahwah, NJ: Lawrence Erlbaum Associates, Inc.

Vellutino, F. R., Tunmer, W. E., Jaccard, J. J., & Chen, R. (2007). Components of reading ability: Multivariate evidence for a convergent skills model of reading development. *Scientific Studies of Reading, 11*, 3–32.

Wechsler, D. (1990). *Wechsler Objective Reading Dimensions (WORD) Pearson assessment*. London: The Psychological Corporation.

Wechsler, D. (1999). *Wechsler Abbreviated Scale of Intelligence (WASI) Harcourt assessment*. San Antonio: The Psychological Corporation.

Wechsler, D. (2001). *Wechsler individual attainment test* (2nd ed.). San Antonio, TX: Pearson Education.

Weekes, B. S., Hamilton, S., Oakhill, J. V., & Holliday, R. E. (2008). False recollection in children with reading comprehension difficulties. *Cognition, 106*, 222.

Wiederholt, J. L., & Bryant, B. R. (2001). *GORT 4: Gray oral reading tests examiner's manual*. Austin, TX: PRO-ED.

Wigfield, A., & Guthrie, J. T. (1997). Relations of children's motivation for reading to the amount and breadth of their reading. *Journal of Educational Psychology, 89*, 420–432.

Yuill, N. (2007). Visiting Joke City: How can talking about jokes foster metalinguistic awareness in poor comprehenders? In D. McNamara (Ed.), *Reading comprehension strategies: Theories, interventions, and technologies* (pp. 325–346). London: Routledge.

Yuill, N., & Joscelyne, T. (1988). Effect of organizational cues and strategies on good and poor comprehenders' story understanding. *Journal of Educational Psychology, 80*, 152–158.

Yuill, N., & Oakhill, J. V. (1988). Effects of inference training on poor reading comprehension. *Applied Cognitive Psychology, 2*, 33–45.

Yuill, N., Oakhill, J. V., & Parkin, A. (1989). Working memory, comprehension ability and the resolution of text anomaly. *British Journal of Psychology, 80*, 351–361.

READING AS AN INTERVENTION FOR VOCABULARY, SHORT-TERM MEMORY AND SPEECH DEVELOPMENT OF SCHOOL-AGED CHILDREN WITH DOWN SYNDROME: A REVIEW OF THE EVIDENCE

Glynis Laws

DEPARTMENT OF EXPERIMENTAL PSYCHOLOGY, UNIVERSITY OF BRISTOL, BRISTOL, UNITED KINGDOM

I. Introduction

A. OUTLINE OF THE QUESTION

Down syndrome (DS) is the most common biological cause of intellectual disability. It is associated with a number of additional impairments including delays and difficulties in the development of language, verbal

Advances in Child Development and Behavior
Patricia Bauer : Editor

short-term memory (VSTM), and speech production. For some years, evidence from case studies and anecdotal reports have suggested that teaching reading to children with DS will support these functions (e.g., Buckley & Bird, 1993; Oelwein, 1995) and, more lately, reading has been recommended as the most effective way to improve them (e.g., Buckley, 2000). Establishing the effectiveness of interventions designed to reduce children's difficulties is important for informing therapy decisions and to meet British Government guidelines that "comprehensive, unbiased and evidence-based information" should be provided to families of children with disabilities (Early Support, 2006). Learning to read provides a number of benefits regardless of whether it improves other functions, and it is within the capabilities of many children with DS. However, the impact of reading as a therapy for oral language, speech, and memory development deserves careful investigation. This chapter reviews evidence from the research literature, and uses data gathered for a recent longitudinal study of reading development in primary school-aged children with DS, to investigate whether reading leads to better receptive vocabulary, VSTM, and speech.

B. DOWN SYNDROME

DS, or trisomy 21, affects around one in 1000 live births in the UK (Morris & Alberman, 2009). The majority of cases are due to nondisjunction of chromosome 21 at meiosis, resulting in three rather than two copies of chromosome 21 in the cells of affected individuals. A small percentage of cases (2–4%) are attributed to translocation of a duplicated section of chromosome 21 to another chromosome. In a similar percentage of cases, nondisjunction during embryogenesis results in a mosaic condition; the proportion of trisomic cells in mosaicism depends on the stage of cell division at which this event occurred.

The DS phenotype is associated with a higher risk of some medical conditions than exists in the general population, including congenital heart disease, hearing loss, leukemia, thyroid disorder, and an early onset of Alzheimer's disease. It is assumed that the phenotypic characteristics of DS are the result of excess genetic material but the ways that genes influence the development of the phenotype are not well understood (Antonarakis & Epstein, 2006; Nadel, 2003; Patterson, 2009). Advances in understanding human genetic disease and disorder, along with the development of mouse models of trisomy 21 to facilitate the search for pre- and postnatal treatments, raise optimism for future therapies to reduce some of the negative effects of DS (Patterson, 2009; Reeves & Garner, 2007).

II. Cognitive Profile in Down Syndrome

Although there is wide individual variation in the presence of many features associated with DS, virtually all individuals have moderate to severe intellectual impairments with IQs in the range of 30–70 (Chapman & Hesketh, 2000; Glenn & Cunningham, 2005). Children with DS progress on tests of nonverbal cognitive ability but IQ scores tend to diminish with chronological age (CA) because scores are indexed relative to those achieved by the typically developing children that provided the test norms. Conversion of raw scores to nonverbal mental ages (MAs) provides a guide to children's level of function but significant floor effects on cognitive tests mean that many children achieve scores that are outside the range of test conversion tables and research reports are often limited to using raw scores. Where studies have reported nonverbal MAs of primary school-aged children with DS, these range from around 2½ to 5½ years (e.g., Caselli, Monaco, Trasciani, & Vicari, 2008). Studies including adolescents and/or young adults report MAs ranging from 4 to 8 years with means of around 5 years (e.g., Glenn & Cunningham, 2005; Kay-Raining Bird, Cleave, White, Pike, & Helmkay, 2008; Laws & Bishop, 2003).

Researchers are some way from understanding how intellectual impairments are related to brain development and function in DS. Although there appear to be few brain differences at birth, structural and functional abnormalities are evident from early postnatal development (see Nadel, 2003). These include delays and differences in the development of the auditory system (e.g., Jiang, Wu, & Liu, 1990; Kittler *et al.*, 2009) and it is plausible that these have a bearing on later hearing difficulties and the speech and language delays associated with DS. A study of auditory event-related potentials in children with DS found differences and variation in the lateralization of processing of speech and nonspeech sounds. However, these differences were not associated with variation in participants' speech and language abilities (Groen, Alku, & Bishop, 2008).

Structural differences in the brains of children and adolescents with DS include reduced volumes of prefrontal cortex, hippocampus, and cerebellum (Jernigan, Bellugi, Sowell, Doherty, & Hesselink, 1993; Pinter, Eliez, Schmitt, Capone, & Reiss, 2001). Pennington, Moon, Edgin, Stedron, and Nadel (2003) have investigated possible links between these brain structures and neuropsychological functions in school-aged children with DS. They have described a weakness in hippocampal function based on deficits in the long-term storage and retrieval of information but found no difference from MA controls on the working memory tasks used to assess the function of prefrontal cortex.

For now, there is limited understanding of the links between brain and cognition, but more is known about relative strengths and weaknesses across the cognitive–linguistic profile in DS. Some aspects of this profile are of particular relevance to the question addressed in this chapter. First, language is delayed relative to nonverbal cognitive function (e.g., Chapman, Schwartz, & Kay-Raining Bird, 1991; Chapman, Seung, Schwartz, & Kay-Raining Bird, 1998) and a modular approach to study has revealed a distinctive profile across the components of the language system. In addition to language impairments, many individuals with DS also have considerable speech deficits. Second, there is dissociation between visuo-spatial short-term memory (VSSTM) and VSTM leading to a well-replicated specific deficit in VSTM.

Researchers have come to understand and describe reading as a process that is "parasitic" on language functions that are impaired in DS. Snowling and Hulme (2005) describe reading comprehension as parasitic on semantics (vocabulary), grammar, and pragmatics, while word reading ability is parasitic on phonology. Thus, the idea of reading as an effective intervention for the deficits associated with DS appears somewhat paradoxical. However, there are plausible ways in which these relationships could operate reciprocally. Features of language, speech, and VSTM functions in DS are described below along with the hypotheses about how these might be improved by reading.

III. Language, VSTM, Speech Production, and Hearing

A. LANGUAGE PROFILE

Reviewers have drawn attention to substantial variation in the severity of language impairments in individuals with DS (e.g., Abbeduto, Warren, & Conners, 2007; Chapman, 1995, 1997, 1999; Chapman & Hesketh, 2000; Fowler, 1990; Laws & Bishop, 2004; Martin, Klusek, Estigarribia, & Roberts, 2009; Roberts, Price, & Malkin, 2007; Rondal, 1995; Tager-Flusberg, 1999; Ypsilanti & Grouios, 2008). A small number of case reports describe individuals who demonstrate exceptional use of language, beyond that predicted by other cognitive functions (Rondal, 1995). However, Fowler (1990) concludes that the expressive use of language is rarely more advanced than that of typically developing 3-year-old children. From an early age, language production develops more slowly than language comprehension (Miller, 1999), and expressive language abilities continue to fall behind nonverbal cognitive abilities throughout childhood and adolescence (Chapman *et al.*, 1991, 1998). Language comprehension is less compromised but there are marked differences between understanding vocabulary and understanding grammar. There are deficits in the use and understanding of morphology (e.g., Eadie, Fey, Douglas, & Parsons, 2002), and numerous studies describe syntax as a specific developmental weakness (e.g., Abbeduto *et al.*, 2003; Chapman *et al.*, 1991; Fowler, 1990; Fowler, Doherty, & Boynton, 1995; Gunn & Crombie, 1996; Laws & Bishop, 2003; Rosin, Swift, Bless, & Vetter, 1988; Tager-Flusberg, 1999). However, receptive vocabulary is reported as a relative strength, either in line with nonverbal MA (Fowler *et al.*, 1995; Laws & Bishop, 2003; Miller, 1999; Vicari, Caselli, & Tonucci, 2000) or advanced relative to nonverbal MA (Barrett & Diniz, 1989; Chapman, 1995; Glenn & Cunningham, 2005; Rosin *et al.*, 1988).

Reading appears to offer scope for increasing vocabulary size. New vocabulary might be introduced explicitly as printed words, or reading could support children's acquisition of vocabulary through the incidental learning of word meanings from context (Nagy, Herman, & Anderson, 1985). Stanovich (1986) has described how a reciprocal relationship between vocabulary and reading accelerates development in typically developing children. Children with better vocabulary are generally more successful readers, making them likely to read more. Exposure to more reading material enables them to acquire more new words, and so they become even more successful as readers, promoting yet further development. It is not clear whether such "bootstrapping" applies to the development of vocabulary and reading in children with DS but, in theory, this appears to be one way in which reading could advance vocabulary knowledge.

B. SPEECH DEFICITS

Many children with DS have speech production difficulties in addition to language impairments (e.g., Cleland, Wood, Hardcastle, Wishart, & Timmins, 2010). As a result, speech can be unintelligible or sometimes intelligible only to close family members. The nature of the speech disorder associated with DS is not well understood and there is debate about whether it results from delayed development of articulation skills (e.g., van Borsel, 1996) or whether the inconsistencies in children's productions of speech sounds are due to poor representation of phonology (Dodd & Thompson, 2001). Cleland *et al.* found no correlations between speech measures and children's IQs or their expressive and receptive language scores, suggesting that the speech disorder is not simply the result of cognitive or language delay. Reading offers the opportunity to practice saying words and it has been suggested that such practice will improve the clarity of children's speech (Bird, Beadman, & Buckley, 2001).

C. DEFICITS IN VERBAL SHORT-TERM MEMORY

Baddeley and Hitch (1974) described the working memory model to conceptualize the processing of information in VSTM and VSSTM. The model specifies a phonological loop, specialized for storage and maintenance of verbal information, and a visuo-spatial sketchpad, specialized for processing visual or spatial material. VSTM is often assessed by digit span, a task that requires participants to remember and repeat progressively longer sequences of digits. VSSTM is assessed by an analogous test, the Corsi blocks task, that requires participants to watch the examiner point to a sequence of locations on a board and then to copy the sequences. Differential performance on these tasks provides evidence for a specific deficit in VSTM in DS (see Jarrold, Purser, & Brock, 2006). VSTM is thought to have an important role in language learning (e.g., Baddeley, Gathercole, & Papagno, 1998). It is also important for reading acquisition by typically developing children (e.g., Gathercole & Baddeley, 1994) and is a correlate of reading ability in individuals with DS (Fowler *et al.*, 1995). Notably, for addressing the principal question of interest here, the relationships between VSTM and reading functions may be reciprocal. For example, Ellis (1990) studied typically developing children between the ages of 5 and 7 years and found that VSTM determined a child's reading success when they first started to read. Once they were making progress, early reading scores were more predictive of later VSTM than were earlier VSTM scores. If the developmental course followed by children with DS is similar to the course followed by the children Ellis studied, then improvements to VSTM in association with reading progress would be a reasonable expectation.

Buckley and Bird (2001) suggest that the auditory training experiences provided by reading will improve the accuracy of representations of the sound patterns of words in the phonological loop. The quality of phonological representations is important for VSTM, notably for completing nonword repetition tasks that depend on adequate phonological storage (see Gathercole, 2006). However, others have argued that nonword repetition depends not only on phonological storage but also reflects phonological processing skills such as segmentation and articulation assembly (e.g., Bowey, 1996, 2001; Snowling, Chiat, & Hulme, 1991). These skills contribute to phonological awareness (PA), the ability to reflect on, and manipulate, the sound structure of language. Thus, phonological memory and PA both may depend ultimately on the clarity or distinctness of underlying phonological representations of speech sounds (Bowey, 1996; Elbro, 1996). Elbro suggests that learning to read is the most effective way to become aware of phonemes. Although children who begin reading with PA are more successful as early readers, reading progress also drives the further development of PA (Perfetti, Beck, Bell, & Hughes, 1987). A slightly less direct link between reading and improved VSTM would take account of lexical restructuring (Fowler, 1991; Metsala, 1999). Increases in oral vocabulary shift the representation of lexical knowledge from whole words to representation of word segments or phonemes since this provides a more efficient way to store and use lexical knowledge. Segmental, or phonemic, representations facilitate the development of PA and of phonological memory (Bowey, 1996). In this view, reading would have a more indirect effect because changes in these functions would depend on reading first leading to increased vocabulary knowledge before lexical restructuring could produce the segmental and phonemic representations necessary to support PA or phonological memory.

In sum, DS is associated with language, VSTM, and speech production difficulties that might all be expected to influence word level reading but for which there are plausible reasons to hope that reading could provide an effective intervention. Before describing features of reading development in DS, a note on hearing loss follows. Hearing impairments are so widespread in DS that it is vital to consider their impact before determining whether the advice to teaching reading as a way to improve other functions is relevant to every person with DS.

D. HEARING LOSS

Research studies record that 40–80% of individuals with DS have hearing impairments (for review, see Davies, 1996; Dahle & McCollister, 1986; Roizen, 1997). In young children, this is largely due to conductive hearing loss associated with persistent glue ear, that is, the sticky effusion that accompanies ear infections. In around 10–15% of individuals, hearing impairments are more severe and aids may be fitted from an early age. There is a trend for hearing to worsen with CA and, from adolescence onward, for sensori-neural hearing losses to develop, and some older individuals are diagnosed with a combination of conductive and sensori-neural losses (Davies, 1996).

Hearing status should be taken into account when drawing conclusions about the effects of DS on development, particularly when considering features such as speech production, language and, possibly, reading that could be vulnerable to hearing loss. The research literature is inconclusive about the functional relationship between hearing and language in DS. Where the primary focus of research has been the investigation of syndrome-specific effects on language development researchers have, quite reasonably, excluded children with more than mild hearing loss to avoid confounding the effects of the syndrome with those of poor hearing (e.g., Abbeduto *et al.*, 2003; Chapman *et al.*, 1991, 1998). Selective samples either report no significant correlations between hearing and language measures (e.g., Abbeduto *et al.*, 2003) or that hearing predicts a small percentage of the variance on tests (e.g., Chapman *et al.*, 1991, 1998). Excluding participants with more severe hearing loss helps to establish the effects on language attributable to DS but other consequences of selective samples are that these may not reflect the full range of language abilities in DS and that the impact of hearing loss on the population with DS may be underestimated.

More marked effects of hearing loss are apparent in unselected samples. Laws (2004) described the expressive language abilities of 30 adolescents and young adults with DS. Mean length of utterance (MLU) provides a guide to expressive language complexity and was derived from narratives that participants provided to accompany a wordless picture book. There was no statistically significant correlation between hearing thresholds and MLU within the group of individuals that had been able to complete the narrative task. In this respect, the findings were similar to those of other studies. However, these individuals either had good hearing or their hearing losses were corrected by aids. The results from seven participants had been excluded from the analysis because they had been unable to provide a narrative. All were found to have moderate hearing losses, but below the threshold where aids would normally be provided. Although it was not possible to make a definite connection between their hearing status and poor expressive language skills, it seems possible that these two functions were related. The effect of poor hearing may not be limited to expressive language; other studies have reported significant effects on receptive vocabulary. Laws and Gunn (2004) noted a significant correlation of −.49 between average hearing threshold and receptive vocabulary scores, after accounting for CA differences in the sample. Jarrold and Baddeley (1997) used a speech discrimination task as an indirect measure of hearing and reported a significant correlation with receptive vocabulary test scores of .64, although no account had been taken of the contribution of CA in this analysis.

Given the additional challenge to language learning posed by hearing difficulties, and the sizeable proportion of children with DS who have hearing impairments, it is important to establish whether the advice to teach reading as the most effective means to improve language applies also to children with DS with hearing losses. This question is of particular interest, given suggestions that children with DS rely on visual learning strategies to acquire literacy and that these strategies allow them to succeed as readers in the face of poor VSTM and low PA (Buckley, 1985; Fidler, Most, & Guiberson, 2005). Thus, it might be reasonable to hypothesize that reading could also provide a visual route into language development for children with poor hearing.

IV. Reading in Children with DS

A. WHAT PROPORTION OF CHILDREN WITH DOWN SYNDROME LEARNS TO READ?

If reading is advised as the most effective way to improve other functions, it is important to know whether this advice applies to all children or whether there are children with DS for whom reading is not a realistic goal. Low levels of reading proficiency have been described historically but opportunities for literacy were more limited in the past (see Buckley, Bird, & Byrne, 1996). Despite recent research interest in reading in DS, and wider provision of literacy teaching, there are few up-to-date reports of literacy rates from samples where all the children have had similar opportunities and/or where the participants have not been preselected as readers. Fidler *et al.* (2005) reported that 65.5% of 7–21-year-olds had some level of reading ability. Participants had not been preselected as readers for this study but children with particularly low verbal abilities had been excluded. If verbal skills are important for acquiring literacy then the results of the study would have overestimated the reading potential of this population. However, the wide age-range could have led to an underestimation since the older participants might have had fewer reading opportunities. Laws and Gunn (2002) reported that only 53% of individuals with DS aged 10–24 years could read one or more words on a standardized test. This group included some older individuals but it was also the case that all the participants were educated in schools for children with special education needs where academic skills were not the main focus of teaching. A more up-to-date guide to literacy rate may be provided by Byrne, MacDonald, and Buckley (2002) in a study of reading by children with DS, all of whom attended mainstream schools, and all of whom had received instruction from the age of 4 years. Eighty-seven-and-a-half percent of these children read at least one word on a standardized reading test by the end of a 2-year study when they had an average age of 10 years 1 month. Similar percentages of participants reached this level of reading in a report by Baylis (2005) and in a recent study of mainstream primary school children to be described in more detail below (Laws, Fisher, Guillaume, Lombard, & Nye, 2010).

B. READING STRATEGIES IN DOWN SYNDROME

Theories of reading development describe the stages through which typically developing readers progress from the early stages of literacy to a skilled level (e.g., Frith, 1985). At first, readers rely on visual recognition of the whole written word and link this to its pronunciation; this is the logographic stage. This strategy depends on learning every printed word along with its spoken form. Later, children reach the alphabetic stage of reading and are able to "sound out" new words based on a knowledge of letter sounds. This phonological route to reading offers a more flexible strategy enabling children to read new words and nonsense words. Skilled readers no longer need to read most words letter-by-letter but recognize them directly. This visual route to reading is also important for reading irregular words, such as yacht, where a phonological approach would not be helpful.

The phonological route to reading depends on the development of PA. PA is established as a predictor of typically developing children's reading success (e.g., Bradley & Bryant, 1983). It has been assessed using a wide range of tasks including letter sound knowledge, and those that require segmentation or blending of words into or from constituent phonemes or larger segments such as rhyme endings. Early reports described children with DS as logographic readers, developing substantial sight word vocabularies (Buckley, 1985), and suggested that they learned to read without PA (Cossu, Rossini, & Marshall, 1993). However, numerous subsequent studies have shown that some individuals with DS do develop PA (for review, see Lemons & Fuchs, 2010). Group performance in these studies is generally significantly poorer than that of comparison groups of typically developing children matched on any criterion. There may also be differences in the order in which particular PA skills are acquired with children more successful on tasks that require detection of onset phonemes than those that depend on sensitivity to rhyme (e.g., Snowling, Hulme, & Mercer, 2002). Despite lower levels of PA, there are similar relationships between PA scores and word reading to those found in MA- or reading-matched groups (e.g., Fowler *et al.*, 1995; Roch & Jarrold, 2008). As in typical development, longitudinal studies show that early PA is predictive of later reading ability (Cupples & Iacono, 2000; Kay-Raining Bird, Cleave, & McConnell, 2000; Laws & Gunn, 2002). In our own recent study, Time 1 PA scores were related to later reading scores but there was no difference between the size of the correlations between regular and irregular words and PA, and no significant correlation between PA and nonword reading. These results suggest that children were not relying on phonological strategies to advance their reading. Interestingly, early

PA was associated with later scores on tests of receptive vocabulary, VSTM, and speech articulation; partial correlations between PA and these functions remained significant after accounting for differences in CA and nonverbal MA, and even early reading scores (Laws *et al.*, 2010).

Despite the importance of PA and letter sound knowledge for facilitating the more flexible phonological reading route, and advice to teach phonics to children with DS once they have established a sight word vocabulary (Buckley *et al.*, 1996), relatively few intervention studies have been published (Cupples & Iacono, 2002; Goetz *et al.*, 2008; Kennedy & Flynn, 2003; van Bysterveldt, Gillon, & Moran, 2006). These studies suggest that improvements to letter sound knowledge, and other PA skills such as onset and rhyme segmentation and blending, are possible following explicit instruction (e.g., Kennedy & Flynn, 2003; van Bysterveldt *et al.*). However, there is less evidence for improvements in word or nonword reading following letter sound and PA instruction (Cupples & Iacono, 2002) suggesting that children may not necessarily use the PA skills they acquire.

For the most part, intervention studies have provided relatively short periods of teaching with sessions totaling 4–8 h spread over 4–6 weeks. Goetz *et al.* (2008) offered a more substantial intervention involving daily 40-min sessions of one-to-one teaching by specially trained learning support assistants over 16 weeks for one group and 8 weeks for a second group that also provided a waiting list control. The first group made gains in letter sound knowledge and early word reading after 8 weeks, relative to the controls. The second group also made small gains after 8 weeks. These gains were statistically nonsignificant but it was thought that this was probably due to a combination of small sample size and wide variation in participants' abilities. The gains made by both groups were maintained after several weeks suggesting that intervention can be successful with higher intensity of teaching over longer periods.

C. IS READING AN EFFECTIVE INTERVENTION FOR LANGUAGE, VSTM, AND SPEECH?

1. A Review

Establishing whether children with DS who learn to read benefit from improved oral language, VSTM, and speech skills is not straightforward. Laws, Buckley, MacDonald, and Broadley (1995) found that the vocabulary and grammar scores of readers and nonreaders with DS diverged over the course of a longitudinal study. The children were aged between 8 years, 8 months and 10 years, 2 months at the beginning of the study, at which time the language scores of the readers did not differ significantly from those of nonreaders. By the end of 4 years, there were statistically significant group differences in their mean vocabulary and grammar scores and in the performance of serial recall tasks. However, these advantages could have been due not to reading *per se* but to the better language models available to the children in the reading group. All of the readers attended mainstream schools whereas the nonreaders attended special schools where they would not have had the benefit of the language environment provided by typically developing classmates. This interpretation is consistent with the results of other studies that have compared placement outcomes, and which suggest that language and VSTM scores of children with DS in mainstream schools are higher than those educated in special schools (Bochner, Outhred, & Pieterse, 2001; Buckley, Bird, Sacks, & Archer, 2006; Laws, Byrne, & Buckley, 2000).

A similar difficulty arises with interpreting the results of studies where some individuals have been brought up at home and others have grown up in institutions, as happened more often in the past (e.g., Bochner *et al.*, 2001). Bochner *et al.* studied language and literacy outcomes for 30 adults with DS and reported that the two individuals with the most advanced vocabulary and reading had been enrolled in a reading intervention program as preschool children. However, Bochner *et al.* also reported language advantages for individuals who had received integrated rather than segregated education and for those brought up at home rather than in institutions, and it is not clear which factors made the greater contribution to the higher functions of the two most advanced adults. Clearly, any research that compares the language skills of children with differing education placements encounters the problem of disentangling the effects of reading instruction from other aspects of the children's school environments or from individual differences among children that could have influenced placement decisions as well as language abilities.

Comparisons of readers and nonreaders within the same education setting should pose fewer difficulties of interpretation. Laws *et al.* (1995) compared the language and VSTM scores of small groups of readers and nonreaders who were all attending special schools. The readers had better scores than nonreaders, which might suggest that reading had accelerated language and memory development. However, the analysis was based on cross-sectional data, so it was equally plausible that having better language and VSTM skills had scaffolded the children's reading. An alternative explanation is that special school teachers had selected the children with better language skills as ready to accept reading instruction (at the time that the study took place it would not have been routine to offer reading instruction to all children in the schools that participated).

Boudreau (2002) also reported significant associations between reading and language measures for 20 children and adolescents with DS but these relationships appeared to be mediated by CA.

Laws and Gunn's (2002) studied individuals with DS attending special schools over 5 years (a different group from the one studied by Laws *et al.*, 1995). Thirty participants were first assessed (Time 1) when they were aged from 5 to 19 years. Although readers had better receptive vocabulary and grammar scores than nonreaders at Time 1, they made no more language progress over the 5 years of study than the nonreaders. Some of the differences between readers and nonreaders were accounted for by hearing difficulties and lower nonverbal cognitive abilities. Both these factors could have influenced children's reading as well as their oral language skills. Reciprocal relationships between language and reading measures at Times 1 and 2 were investigated using partial correlations, taking test scores at Time 1 into account. Receptive vocabulary and grammar scores at Time 1 predicted reading scores at Time 2, but reading at Time 1 did not predict receptive vocabulary and grammar at Time 2. Time 1 reading did predict MLU, a measure of expressive language complexity, at Time 2. However, since MLU had not been measured at Time 1, it had not been possible to take expressive language abilities at the outset into account in the partial correlation used for this investigation. Sentence recall provides an alternative measure of children's expressive language. Sentence recall had been measured at both time points and, once scores for sentence recall at Time 1 as well as CA were taken into account, reading scores predicted neither sentence recall nor MLU at Time 2 (unpublished analysis).

Laws and Gunn's (2002) results were based on longitudinal data from individuals with similar special education placements, thus overcoming some of the problems of interpretation when comparing children in different placements. Despite this, the evidence that the study offered against the idea that reading precedes better language development was rather weak. The participants were drawn from a wide CA range, which is not unusual in DS research, but the language and memory scores of some of the older participants had declined over the 5 years of study (Laws & Gunn, 2004). The reason for the decline was not clear but, whatever its cause, the negative change in scores is likely to have affected the results of the analysis. Also, reading inputs were not controlled and it is not known for how long the participants had been receiving instruction. For example, five children only emerged as readers at the end of the study when they had an average CA of 12 years 10 months.

A study by Byrne *et al.* (2002) should provide a clearer guide to the association between reading and language and VSTM since it included children from a more constrained age-range and reading inputs were controlled. These researchers studied 24 children, aged from 4 years, 11 months to 12 years, 7 months, for 2 years. As described above, most of the children achieved a reading score by the end of the study. There were no concurrent correlations between reading and language or VSTM scores once CA was accounted for. No correlations between Time 1 reading and final language measures were reported. Baylis (2005) was able to study the language and reading progress of 38 individuals with DS over the longer period of 3 years, 6 months. At the outset, the children were aged from 7 years to 16 years 6 months and attended mainstream schools, although about one quarter of them had moved to special schools by the end of the study. More proficient readers made more progress than beginning readers on a composite language measure derived from receptive vocabulary scores and scores from a narrative task. The groups differed in MLU but there was no significant increase in MLU in the course of the study, and no suggestion that proficient readers had made more progress. Within the whole sample, reading was a significant predictor of the language composite, except when expressive language abilities were controlled.

Although there is limited evidence from group studies to support the idea that reading boosts language and memory in children with DS, if reading does have this effect, we might expect that individuals with exceptionally strong literacy skills would have developed concomitant exceptional oral language abilities. Groen, Laws, Nation, and Bishop (2006) described KS, a girl with DS aged 8 years 2 months. She had exceptional literacy skills compared to those of a group of other readers with DS around the same CA. In particular, she had unusually good nonword reading skills, that is, she was able to use her knowledge of letter sounds to read pseudo-words (e.g., kisp) that she had not previously encountered. However, despite superior reading skills, KS did not have statistically significantly better scores in vocabulary, grammar, sentence recall, or VSSTM tests than the comparison group. KS had a digit span of four items, which was significantly longer than the mean span of just under three items calculated for the comparison group. Her parents completed the Children's Communication Checklist 2 (Bishop, 2003), a questionnaire to rate speech and structural language abilities as well as social communication skills. KS did not differ from the comparison group on any components of this measure although the difference between her scaled score for speech and that of the comparison group mean approached significance. KS achieved language and memory scores within the upper quartiles of the ranges for children in the comparison group, although her language scores were not as exceptional, relative to this group, as her reading abilities. However, since the study did not report longitudinal data, it was not possible

to judge whether good language, VSTM and speech production contributed
to her success as a reader or were a consequence of reading.

2. A Study to Investigate the Relationships Between Early Word Recognition and Later Receptive Vocabulary, VSTM and Speech Production

Laws *et al.* (2010) have studied the oral language, memory, and reading
abilities of children with DS who were all attending mainstream primary
schools and receiving literacy instruction. By following the development
of reading, as well as a wide range of other cognitive abilities including language and memory, for 2 or more years, it was possible to investigate the
effects of early reading on the development of other functions. The main
focus of the research had been a comparison of the literacy development
of children with DS to that of children with specific language impairment
(SLI). There are similarities between the language and VSTM profiles of
the two populations (Eadie *et al.*, 2002; Laws & Bishop, 2003, 2004) but it
is not clear whether reading follows similar developmental pathways.
Whereas numerous studies point to persistent reading delays and difficulties
for individuals with SLI (e.g., Bishop & Adams, 1990; Snowling, Bishop, &
Stothard, 2000, Whitehouse, Line, Watt, & Bishop, 2009), reading has often
been reported as a developmental strength in DS (e.g., Buckley *et al.*, 1996).
The first aim of Laws *et al.*'s research had been to compare the two groups to
investigate whether there were differences between them in the ways that
literacy is acquired, and to examine the reading strategies used by the children. Findings were very similar for both groups: early letter knowledge,
PA, and speech accuracy were correlated with final word reading scores
after accounting for CA. In the group with DS, expressive and receptive
vocabulary were also associated with final reading scores, and in the group
with SLI, VSTM was related. A secondary aim had been to investigate evidence to support the claim that teaching reading improves the oral language,
VSTM, and speech production of children with DS. The remainder of this
chapter focuses on this second aim and describes only the study results
obtained from the participants with DS.

Twenty-eight children (10 boys, 18 girls) took part, aged from 5 to 11
years at the start of the study (Time 1). Most families reported a diagnosis
of trisomy 21 but two girls had mosaicism, a diagnosis associated with variable outcomes in terms of the nature and severity of associated
impairments (Papavassiliou *et al.*, 2009). For one girl (CA = 5 years,
9 months), IQ and other test scores were within the ranges of the scores
achieved by other children of similar CA in the sample. The other girl
(CA = 7 years, 10 months) also had an IQ within the sample range but

her reading, vocabulary, and speech production scores were exceptionally high. The scores of both children were retained in the analyses. All the children were receiving literacy instruction in mainstream primary schools as part of the regular curriculum.

The children completed a battery of assessments for the project of which a subset is described here. Nonverbal cognitive ability was assessed using the Leiter International Performance Scale—Revised (Leiter-R) (Roid, Miller, & Lucy, 1997). The Leiter-R minimizes the influence of limited linguistic skills on performance by the use of pantomime in its administration and the requirement for nonverbal responses. Four subtests provided a brief nonverbal IQ and raw scores were converted to MAs (Leiter MAs). Receptive vocabulary was assessed by the British Picture Vocabulary Scale II (BPVS II) (Dunn, Dunn, Whetton, & Burley, 1997). The Goldman Fristoe Test of Articulation, second edition (GF2) (Goldman & Fristoe, 2000) provided a guide to speech production abilities. The test requires children to name pictures and the responses are used to assess whether specific consonants or consonant clusters have been correctly articulated. The percentage of consonants and consonant clusters produced correctly was used in the analyses. A digit span test provided an assessment of children's VSTM. Digit scores were reported as the number of test items recalled correctly while digit spans recorded the longest sequence of numbers recalled in the correct order.

Children's reading abilities were assessed using the word reading subtest from the British Ability Scales II (BAS II) (Elliott, Smith, & McCulloch, 1996). The test provides a guide to children's word recognition abilities. All the children were tested from the beginning of the list of 90 test words. Raw scores were used in the statistical analyses and scores were also converted to RAs to describe the levels of reading function. Parents provided details about current and past hearing difficulties and treatments, and gave consent for children's audiology clinics to be contacted for records of hearing tests.

Assessments were administered at the beginning of the project (Time 1) and readministered after an average interval of 27 months (Time 2). Table I shows CAs and all test scores at both assessment points. Intervals between assessment points varied (range = 18–31 months) because recruitment had been slow and it had not been possible to allow the same interval for some late recruits as provided for earlier recruits within the period of funding. Most children were assessed in school over two or three visits at each time point but two children were assessed at home at the request of their parents.

At Time 1, 75% (21/28) children achieved a raw score of at least one point on the BAS II word reading test. By Time 2, 93% (26/28) of the children achieved one point or more and RAs ranged from 5 years to

Table I
Mean (SD), CA, Leiter MA and Test Scores at Time 1 and Time 2 ($N=28$)

	Time 1	Time 2
CA (months)	92.75 (20.86)	118.29 (21.03)
Leiter MA (months)	47.21 (8.80)	52.00 (8.13)
BAS II raw score	12.71 (15.67)	23.82 (19.88)
BAS II RA (months)	69.86 (12.86)	78.61 (17.05)
BPVS II raw score	37.41 (15.24)	44.54 (17.83)
BPVS II AE[a] (months)	48.38 (13.57)[b]	57.40 (16.56)[c]
Digit span	2.18 (1.39)	2.96 (.81)
Digit score	3.04 (2.08)	4.64 (2.08)
Goldman Fristoe[e]	52.41 (28.34)[d]	59.59 (27.03)[d]

[a]Age equivalent.
[b]$N=26$ as two raw scores were out of range of norms.
[c]$N=25$ as three raw scores were out of range of norms.
[d]$N=27$ as one child was not tested.
[e]Percentage of sounds correctly articulated.

10 years, 3 months. Reading progress was variable with increases in RAs from Time 1 to Time 2 ranging from 2 months to 2 years, but the gap between RA and CA widened for every child. The median difference between RA and CA by Time 2 was 3 years, 4 months (excluding the children that scored zero). The RAs of all the children were higher than their Leiter MAs and, except for one child, RAs were also higher than the age equivalents for scores obtained on the BPVS II. Three children were exceptionally good readers with BAS II standard scores of more than 90 (mean=94, SD=3). Figure 1 shows the distributions of reading raw scores at Times 1 and 2.

Children were assigned to one of two groups on the basis of reading scores at Time 2: a "low" group, including those that scored zero or had RAs of 5 years, 10 months or less; and a "high" group, including those with RAs of 6 years, 4 months or more. If reading has a positive impact on other functions, children in the "high" reading group would be expected to have significantly better language, VSTM, and speech production. Mixed analyses of variance (ANOVAs) were used to investigate this hypothesis. A significant main effect of the between-groups factor would suggest a relationship between reading and, for example, vocabulary, but would provide no information about the direction of this relationship. An indication that being a good reader led to improvements in other functions would be a widening gap between the scores for the two groups by Time 2, which would be indicated by a significant statistical interaction between the between-groups and the repeated measures factors.

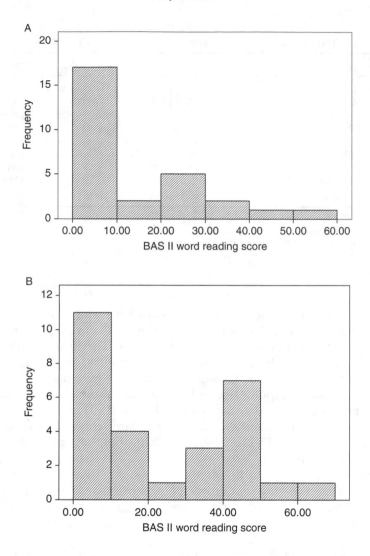

Fig. 1. (A) Distribution of raw scores on BAS II word reading test at Time 1. (B) Distribution of raw scores on BAS II word reading test at Time 2.

Table II shows the distribution of Time 1 and Time 2 reading scores for the "low" and "high" groups. The "high" group was significantly older than the "low" group and the groups also differed in terms of Leiter MA, although this difference was partly accounted for by CA differences. CA and Leiter MA made independent contributions to one or more test scores.

Table II

Mean (SD), CA, Leiter MA and Test Scores at Time 1 and Time 2 for "Low" and "High" Reading Groups

	"Low" Readers ($N=13$)		"High" Readers ($N=15$)	
	Time 1	Time 2	Time 1	Time 2
CA (months)	83.08 (15.75)	108.62 (17.65)	101.13 (21.5)	126.67 (20.59)
Leiter MA (months)	42.62 (7.92)	46.85 (6.69)	51.20 (7.66)	56.47 (6.69)
BAS II raw score	1.69 (2.39)	5.77 (4.30)	22.26 (16.06)	39.47 (13.47)
BAS II RA (months)	60.85 (1.99)[a]	63.31 (3.77)[a]	77.67 (13.22)	91.87 (11.95)
BPVS II raw score	25.85 (11.09)	33.62 (17.09)	46.93 (11.01)	54.00 (12.50)
BPVS II AE (months)	38.36 (6.38)[b]	47.80(16.19)	55.73 (12.78)	63.80 (13.85)
Digit span	1.38 (1.39)	2.54 (.66)	2.87 (.99)	3.33 (.82)
Digit score	1.92 (2.06)	3.77 (1.83)	4.00 (1.60)	5.40 (2.03)
Goldman Fristoe[e]	37.54 (31.97)	45.0 (31.0)[c]	66.21 (17.17)[d]	71.67 (16.35)

[a]Children scoring zero were awarded an RA of 60 months.
[b]$N=11$ as two raw scores out of range of test norms.
[c]$N=12$ as one child not tested.
[d]$N=14$ as one child not tested.
[e]Percentage of sounds correctly articulated.

Neither CA nor Leiter MA was significantly correlated with reading scores but there were trends in this direction. Given these associations, CA and Leiter MA were included as covariates in the statistical analyses.

There were no significant differences between girls' and boys' CA, Leiter MA, or scores on any of the tests. There was no difference in the length of the interval between assessments for the two groups ("low" group: mean $=26$ months, SD $=4$ months; "high" group: mean $=25$ months, SD $=4$ months). However, across the whole sample, there were significant correlations between interval and the size of the gains recorded for receptive vocabulary and speech production so interval was taken into account in the statistical analyses.

An analysis to investigate changes in receptive vocabulary scores indicated no significant effect of time of assessment, suggesting that there was no overall increase in scores from Time 1 to Time 2. A significant interaction between interval and time of assessment showed that children who had a longer interval between assessments tended to do better at Time 2 than those with shorter intervals. The "high" reading group did have better vocabulary scores than the "low" group but there was no indication that the "high" group had improved differentially since, critically, there was no significant statistical interaction between reading group and time of measurement. The superior vocabulary scores of the "high" reading group were attributable to Leiter MA and CA differences since both these factors contributed to variation in vocabulary scores. A further investigation of scores within the "high" reading group provided no evidence that Time 1 reading was significantly correlated with Time 2 vocabulary, even after accounting for CA and Leiter MA. In sum, despite considerably better reading abilities, there was no evidence that children classified as "high" readers at Time 2 had improved more on the receptive vocabulary test than those classified as "low" readers, and no suggestion that their receptive vocabulary was predicted by early reading scores.

A similar analysis to investigate changes in VSTM showed that "low" and "high" readers did not differ significantly in terms of digit scores. The mean increase in scores between Time 1 and Time 2 was greater for the "low" reading group but there was no significant effect of time of measurement and no interaction between time of measurement and reading group (all $Fs < 1$, ns). In sum, the children in the "high" reading group did not have better VSTM than those in the "low" reading group, and made no more progress.

The percentage of sounds produced correctly on the GF2 test was used to investigate the effect of reading group membership on speech production differences. There was a nonsignificant trend toward readers having higher scores than nonreaders but no significant overall increase in scores, and no interaction between reading group and the repeated measures factor to support the hypothesis that the "high" group would make more progress on the test than the "low" group. In sum, there was no evidence for a link between reading group and improvements in speech production.

As discussed earlier in the chapter, understanding how hearing loss contributes to children's reading and language progress is an important issue. Hearing information was available for 25 children, provided by parents and/or by audiology clinics. This information was used to classify children according to hearing status. One group included children for whom there were no parent-reported difficulties and/or for whom we received audiology test results that showed hearing thresholds of 30 dB or lower at frequencies of .5, 1, 2, and 4 kHz in both ears ($N = 17$). The other group included eight children who either wore hearing aids or who had audiology test results showing hearing thresholds of 35 dB or higher for sounds at one or more of these frequencies. None of these children was in the "high" reading group. Thus, hearing status was an important determinant of whether children with DS in this study acquired good levels of literacy.

The results reported above for the "low" reading group would have been influenced by the inclusion of children with hearing difficulties. Mean test scores were recalculated after excluding these eight children (see Table III). Despite a small, although statistically nonsignificant, mean CA advantage, and significantly better mean Leiter MA and reading scores, the "high" reading group did not do better than the "low" reading group on any of the measures of interest.

Table III

Comparison of CA, Leiter MA and Time 2 Scores for "Low" and "High" Reading Groups
After Exclusion of Children with Hearing Difficulties

	"Low" Readers ($N=5$)	"High" Readers ($N=15$)
CA (months)	120.00 (17.04)	126.67 (20.59)
Leiter MA (months)	46.20 (7.92)[*]	56.47 (6.69)[*]
BAS II raw score	5.80 (5.45)[*]	39.47 (13.47)[*]
BPVS II raw score	47.60 (17.37)	54.00 (12.50)
GF2 % correct	75.00 (13.84)	71.67 (16.35)
Digits score	5.60 (.55)	5.40 (2.03)
Digit span	3.20 (.45)	3.33 (.82)

*t-tests indicate scores significantly different, $p < .05$.

As a final test of the hypothesis that reading would be associated with bet-ter vocabulary, VSTM, and speech production, the gains made by three par-ticularly strong readers were compared with those of the "low" reading group. It is worth noting that the children with hearing difficulties were retained in the "low" group. Three girls had CA-appropriate reading and achieved BAS II reading standard scores > 90 at both assessment points (Time 1: mean $= 103$, SD $= 8$; Time 2: mean $= 94$, SD $= 3$). At Time 2, one girl had a RA $= 7$ years, 1 month, and two girls had RAs $= 8$ years, 3 months (including one of the children with mosaicism). If reading improves the other functions we investigated, it would be reasonable to expect the benefits to be manifested most clearly in comparisons between these chil-dren and those who remained as nonreaders or emerging readers by Time 2. The groups did not differ significantly in CA or Leiter MA, and there were similar intervals between the two assessment points. The strong readers had made considerable reading progress between the two assessments but the groups did not differ significantly in terms of the progress made on the vocabulary, VSTM, or speech articulation tests (see Table IV).

In summary, like other studies (Baylis, 2005; Byrne et al., 2002), this research found that when children with DS have access to mainstream edu-cation and are taught reading as part of the regular curriculum, around 90% of them will learn to read sufficiently well to achieve a score on a standardized test of reading. There was a wide range of reading abilities within the sample, evident from the standards of reading reached and from the amount of progress made over the course of the study. All the children achieved RAs below CAs with a median difference of 3 years 4 months, that is, somewhat larger than the estimate of 2 years provided by Buckley (2001). RAs of readers with DS were in advance of Leiter MAs, and most were in advance of children's BPVS II age equivalent scores. A comparable

Table IV

Comparison of CA, Leiter MA, and Gains Made by "Strong" Readers and "Low" Readers
by Time 2

	"Low" Readers ($N=13$)	"Strong" Readers ($N=3$)
CA (months)	108.62 (17.65)	101.67 (12.06)
Leiter MA (months)	46.85 (6.69)	54.00 (8.72)
BAS II gain	4.08 (3.82)*	19.0 (10.58)*
BPVS II gain	7.77 (13.99)	4.33 (13.28)
GF2 gain	8.92 (15.50)[a]	13.33 (18.18)
Digits score gain	1.85 (1.34)	1.67 (.58)
Digit span gain	1.00 (1.00)	1.00 (1.00)

[a]$N=12$.
*Mann–Whitney test indicates scores significantly different, $p<.01$.

dissociation between RAs and other cognitive tests was described as indicating "strength" in reading in DS (Buckley *et al.*, 1996). However, typically developing children would not normally be taught to read at CAs equivalent to the vocabulary ages and nonverbal MA levels of these children with DS. Further, children who were barely reading, or not reading at all, would have been awarded a RA of 5 years on the BAS word reading test (Elliott, 1983) and, while this is likely to be in advance of other age equivalent scores, it is arguable whether describing this dissociation as illustrative of reading "strength" is useful. It could be more useful to describe reading achievements, not in relation to other functions or to CA, but in terms of absolute level of reading ability and what that provides in terms of access to literature for enjoyment and education.

In the study reported, 43% of the children achieved RAs of 7 years or more with some children (just under 20%) achieving RAs of 8 years or more. These levels of skill should ensure that a wide range of reading material is accessible, although reading comprehension abilities, not dealt with in this review, would be important in determining the real benefits that children derive from this access. Of course, most of the children were still in primary school at Time 2 and hopefully will continue to advance as readers.

Children with poor hearing did least well, either remaining as nonreaders throughout the study or making considerably less progress than most other children. Nonverbal MA was also a significant factor in children's learning to read. There was little evidence that more competent readers developed more in terms of receptive vocabulary, VSTM, or speech production. The three most advanced readers made no more gains in the functions studied than children in the low reading group.

V. Summary and Future Directions

The main question addressed by this chapter was whether there is evidence to support the view that reading provides the most effective intervention for oral language, VSTM, and speech development for individuals with DS. It was assumed that, if this was the case, proficient readers would develop more advanced functions than nonreaders or limited readers. Although a review of the literature showed associations between reading and other functions, there was little clear evidence that reading was the driver in these relationships. In part, this was due to methodological difficulties within most studies that have addressed the question, such as confounds between access to instruction and educational placements, the inclusion in samples of older participants that might not have had opportunities to learn, or the weakness of cross-sectional research designs. The reported study compared the receptive vocabulary, VSTM, and speech production of nonreaders and emerging readers, with RAs of 5 years, 10 months or less, with those of more competent readers, with RAs of 6 years, 4 months or more. There was no evidence that the better readers had improved more in terms of other functions. Perhaps the most telling result was that there was no suggestion that three children who were particularly strong readers had made greater gains on the receptive vocabulary, VSTM, or speech production measures than the low reading group, despite having comparable CAs and MAs, and having made substantial average reading gains of 20 points compared to average gains of less than four points made by the low group.

Given wide acceptance of the view that teaching reading will develop the spoken language skills of children with DS (Alton, 2006; Buckley, 2000; Early Support, 2005; Oelwein, 1995), it is important to consider possible reasons for these negative findings. Buckley (2001) has argued that simple exposure to print, rather than learning to read, is all that is required for children to benefit from written language. In this view, the reason that more proficient readers in the study do not have better language and memory could be because the nonreaders and emerging readers do equally well since all the children have been exposed to print. This seems unlikely; there was wide variation in language scores within both groups and the null results were not simply due to differences in means narrowly missing the level required for statistical significance. For example, at Time 1, one of the highest scoring readers had a receptive vocabulary standard score of 47 at the age of 10 years, while a child of similar CA from the low reading group had a standard score of 74. Of course, it is not possible to know whether both these children have attained higher levels of vocabulary than they would have attained without exposure to print. Teaching reading, particularly in a supportive mainstream environment, provides opportunity to focus on language skills that could benefit children regardless of the level of reading they ultimately achieve. However, whether this is the case would be a difficult research question to address, and is rather different from the question addressed in this chapter.

Buckley (2001) suggests that children who are introduced to reading in their preschool years show the highest levels of achievement, so possibly the school-aged children in the studies reviewed started reading too late to experience a benefit. There is no doubt that many preschool children with DS readily learn to identify words and may make rapid progress (e.g., Appleton, Buckley, & MacDonald, 2002; Buckley & Bird, 1993). Sixty-one percent of preschoolers with DS studied by Appleton *et al.* became readers and, after 3 years, there was a widening gap in expressive and receptive language between these children and those that remained nonreaders, although no statistical analysis was provided to show if differences were significant. More research evidence is needed to decide whether such children ultimately achieve more in terms of language and memory development than children with DS who are not given reading opportunities until later. Such evidence would support the earlier provision of education resources to allow for teaching reading to preschool children with DS (see Bird & Buckley, 2001).

There is also the possibility that children may not have been studied for long enough to observe benefits from reading. Progress in language and memory development is slow in DS and there is poor consolidation of learning; relatively small mean increases in scores on language and cognitive tests were reported by Byrne *et al.* (2002) and Laws *et al.* (2010) after 2 years or so of study. However, although the studies took place over a relatively short period, the children had been taught reading for longer so, if reading does improve other functions, it might be expected that any differences between limited and proficient readers should have had time to emerge. Hopefully, it will be possible to monitor the development of Laws *et al.*'s participants to check whether, over a longer time scale, differential progress between good and poor readers is established, and if evidence emerges for changes in the direction of relationships between oral language, VSTM, and reading development.

One other point to make is that the standardized receptive vocabulary test that children completed provides a guide to the sophistication of vocabulary knowledge rather than an indication of the actual number of words known. Although these two aspects of vocabulary seem likely to be related, it is conceivable that reading could increase the number of simple words known by a child without them necessarily acquiring the more advanced vocabulary necessary to score more on the standardized test. This question was not addressed in the study reported.

In common with earlier studies, nonverbal MA was an important determinant of reading success (Carr, 1995; Laws & Gunn, 2002; Sloper, Cunningham, Turner, & Knussen, 1990). The study also highlighted the issue of hearing difficulties as a factor in reading acquisition. Notably, eight children with hearing problems remained as nonreaders or limited readers throughout the study. Despite the emphasis that has been placed on visual learning for the acquisition of reading skills (e.g., Fidler *et al.*, 2005), it appears that written language may not compensate for the difficulties caused by hearing loss. The importance of regular hearing checks, appropriate treatment for hearing disorders and measures to ensure that children are provided with every chance to hear and participate in the classroom cannot be overemphasized. It is important that behaviors that could be attributed to poor hearing, such as attention difficulties or lack of communication, are not mistakenly attributed to a child having DS. Bird *et al.* (2001) provide guidance for compensating for hearing loss when teaching reading to children with DS. Given the numbers of children with additional hearing problems, and the apparent effect of this on reading, it is important that future research into the effectiveness of interventions examines whether they will help these children as well as those with good or reasonable hearing.

In conclusion, the majority of school age children with DS learn to read words but there is a wide variation in achievements related to hearing status and nonverbal MA. There was little evidence to support the hypothesis that better readers would be more advanced than nonreaders or emerging readers in terms of vocabulary, VSTM, or speech production. Despite this, it is important that parents' and teachers' efforts in providing literacy teaching to children with DS should continue. However, it is also important that any interventions that are recommended are presented with accurate, evidence-based information that fosters realistic expectations as to their outcomes.

Although reading was not associated with improvements in receptive vocabulary or speech development, it is worth drawing attention to the finding that early PA was a significant predictor of these functions as well as of VSTM at Time 2, even after accounting for CA, Leiter MA, and early reading scores (Laws *et al.*, 2010). The intervention provided by Goetz *et al.* (2008) showed that children with DS could be taught PA successfully. This raises the possibility that teaching PA, either independently or as part of an integrated approach to literacy teaching, could have an impact on oral language and memory. Features of Goetz *et al.*'s study that contributed to its success were frequent, regular practice, over an extended period, delivered by trained learning support staff. Further research is required to investigate whether the effects of an intervention to teach PA, incorporating these features, might benefit language and memory development directly, as well as improve literacy skills.

Acknowledgments

The study described in this chapter was funded by a grant from the Wellcome Trust (grant no. 076250). Time 1 assessments and data entry were carried out by Frances Lombard and Joanna Nye, and Time 2 assessments and data entry were carried out by Alison Fisher and Stephanie Guillaume, with contributions from Heather Brown, Beth Main and Catherine White. The work of these research assistants and students is gratefully acknowledged, as is the interest and cooperation of the children that participated in the research and that of their parents, teachers, and teaching assistants.

REFERENCES

Abbeduto, L., Murphy, M. M., Cawthon, S. W., Richmond, E. K., Weissman, M. D., Karadottir, S., *et al.* (2003). Receptive language skills of adolescents and young adults with Down or Fragile X syndrome. *American Journal of Mental Retardation, 108,* 149–160.

Abbeduto, L., Warren, S. F., & Conners, F. A. (2007). Language development in Down syndrome: From the pre-linguistic period to the acquisition of literacy. *Mental Retardation and Developmental Disabilities Research Reviews, 13,* 247–261.

Alton, S. (2006). *Reading—Information sheet.* London: Down Syndrome Association Education Consortium.

Antonarakis, S. E., & Epstein, C. J. (2006). The challenge of Down syndrome. *Trends in Molecular Medicine, 12,* 473–479.

Appleton, M., Buckley, S., & MacDonald, J. (2002). The early reading skills of preschoolers with Down syndrome and their typically developing peers—Findings from recent research. *Down Syndrome News and Update, 2,* 9–10.

Baddeley, A. D., Gathercole, S. E., & Papagno, C. (1998). The phonological loop as a language learning device. *Psychological Review, 105,* 158–173.

Baddeley, A. D., & Hitch, G. (1974). Working memory. In G. Bower (Ed.), *The psychology of learning and motivation* (Vol. 8, pp. 47–90). New York: Academic Press.

Barrett, M. D., & Diniz, F. A. (1989). Lexical development in mentally handicapped children. In M. Beveridge, G. Conti-Ramsden & I. Leudar (Eds.), *Language and communication in mentally handicapped people.* London: Chapman & Hall.

Baylis, P. (2005). *Reading skills in Down syndrome.* University of York, Unpublished PhD thesis.

Bird, G., Beadman, J., & Buckley, S. (2001). *Down syndrome issues and information: Reading and writing for individuals with Down syndrome (5–11 years).* Portsmouth: Down Syndrome Educational Trust.

Bird, G., & Buckley, S. (2001). *Down Syndrome issues and information: Reading and writing development for infants with Down syndrome (0–5 years).* Portsmouth: Down Syndrome Educational Trust.

Bishop, D. V. M. (2003). *Children's communication checklist, version 2.* London: Psychological Corporation.

Bishop, D. V. M., & Adams, C. (1990). A prospective study of the relationship between specific language impairment, phonological disorders and reading retardation. *Journal of Child Psychology and Psychiatry, 31*, 1027–1050.

Bochner, S., Outhred, L., & Pieterse, M. (2001). A study of functional literacy skills in young adults with Down syndrome. *International Journal of Disability, Development and Education, 48*, 67–90.

Boudreau, D. (2002). Literacy skills in children and adolescents with Down syndrome. *Reading and Writing, 15*, 497–525.

Bowey, J. A. (1996). On the association between phonological memory and receptive vocabulary in five year olds. *Journal of Experimental Child Psychology, 63*, 44–78.

Bowey, J. A. (2001). Nonword repetition and young children's vocabulary: A longitudinal study. *Applied Psycholinguistics, 22*, 441–469.

Bradley, L., & Bryant, P. (1983). Categorizing sounds and learning to read: A causal connection. *Nature, 301*, 419.

Buckley, S. J. (1985). Attaining basic educational skills: Reading, writing and number. In D. Lane & B. Stratford (Eds.), *Current approaches to Down's syndrome* (pp. 315–343). London: Holt Saunders.

Buckley, S. J. (2000). *Down syndrome issues and information: Speech, language and communication for individuals with Down syndrome: An overview.* Portsmouth: Down Syndrome Education International.

Buckley, S. J. (2001). *Down syndrome issues and information: Reading and writing for individuals with Down syndrome: An overview.* Portsmouth: Down Syndrome Education International.

Buckley, S. J., & Bird, G. (1993). Teaching children with Down syndrome to read. *Down Syndrome Research and Practice, 1*, 34–39.

Buckley, S. J., & Bird, G. (2001). *Down syndrome issues and information: Memory development for individuals with Down syndrome: An overview.* Portsmouth: Down Syndrome Education International.

Buckley, S., Bird, G., & Byrne, A. (1996). Reading acquisition by young children. In B. Stratford & P. Gunn (Eds.), *New approaches to Down syndrome* (pp. 268–279). London: Cassell.

Buckley, S., Bird, G., Sacks, B., & Archer, T. (2006). A comparison of mainstream and special education for teenagers with Down syndrome: Implications for parents and teachers. *Down Syndrome Research and Practice, 9*, 54–67.

Byrne, A., MacDonald, J., & Buckley, S. (2002). Reading, language and memory skills: A comparative longitudinal study of children with Down syndrome and their mainstream peers. *The British Journal of Educational Psychology, 72*, 513–529.

Carr, J. (1995). *Down's syndrome: Children growing up.* Cambridge: Cambridge University Press.

Caselli, M. C., Monaco, L., Trasciani, M., & Vicari, S. (2008). Language in Italian children with Down syndrome and with specific language impairment. *Neuropsychology, 22*, 27–35.

Chapman, R. S. (1995). Language development in children and adolescents with Down syndrome. In P. Fletcher & B. MacWhinney (Eds.), *Handbook of child language* (pp. 641–663). Oxford: Blackwell Publishers.

Chapman, R. S. (1997). Language development in children and adolescents with Down syndrome. *Mental Retardation and Developmental Disabilities Research Reviews, 3*, 307–312.

Chapman, R. S. (1999). Language development in children and adolescents with Down syndrome. In J. F. Miller, M. Leddy & L. A. Leavitt (Eds.), *Improving the communication of people with Down syndrome* (pp. 41–60). Baltimore, MD: Paul Brookes Publishing.

Chapman, R. S., & Hesketh, L. J. (2000). Behavioural phenotype of individuals with Down syndrome. *Mental Retardation and Developmental Disabilities Research Reviews, 6*, 84–95.

Chapman, R. S., Schwartz, S. E., & Kay-Raining Bird, E. (1991). Language skills of children and adolescents with Down syndrome: I. Comprehension. *Journal of Speech and Hearing Research, 34*, 1106–1120.

Chapman, R. S., Seung, H.-K., Schwartz, S. E., & Kay-Raining Bird, E. (1998). Language skills of children and adults with Down syndrome: Production deficits. *Journal of Speech, Language, and Hearing Research, 41*, 861–873.

Cleland, J., Wood, S., Hardcastle, W., Wishart, J., & Timmins, C. (2010). Relationship between speech, oromotor, language and cognitive abilities in children with Down's syndrome. *International Journal of Language & Communication Disorders, 45*, 83–95.

Cossu, G., Rossini, F., & Marshall, J. C. (1993). When reading is acquired but phonemic awareness is not: A study of literacy in Down's syndrome. *Cognition, 46*, 129–138.

Cupples, L., & Iacono, T. (2000). Phonological awareness and oral reading skill in children with Down syndrome. *Journal of Speech, Language, and Hearing Research, 43*, 595–608.

Cupples, L., & Iacono, T. (2002). The efficacy of 'whole word' versus 'analytic' reading instructions for children with Down syndrome. *Reading and Writing, 15*, 549–574.

Dahle, A. J., & McCollister, F. P. (1986). Hearing and otologic disorders in children with Down syndrome. *American Journal of Mental Deficiency, 90*, 636–642.

Davies, B. (1996). Auditory disorders. In B. Stratford & P. Gunn (Eds.), *New approaches to Down syndrome* (pp. 100–121). London: Cassell.

Dodd, B., & Thompson, L. (2001). Speech disorder in children with Down's syndrome. *Journal of Intellectual Disability Research, 45*, 308–316.

Dunn, L. M., Dunn, L. M., Whetton, C., & Burley, J. (1997). *The British picture vocabulary scale* (2nd ed.). Windsor, UK: NFER-Nelson.

Eadie, P. A., Fey, M. E., Douglas, J. M., & Parsons, C. L. (2002). Profiles of grammatical morphology and sentence imitation in children with specific language impairment and Down syndrome. *Journal of Speech, Language, and Hearing Research, 45*, 720–732.

Early Support, (2005). *Information for parents: Down syndrome*. London: DfES Publications. *http://www.earlysupport.org.uk/*.

Early Support, (2006). *Informed choice, families and deaf children: Professional handbook.* London: DfES Publications.*http://www.earlysupport.org.uk/*.

Elbro, C. (1996). Early linguistic abilities and reading development: A review and a hypothesis. *Reading and Writing, 8*, 453–485.

Elliott, C. D. (1983). *British ability scales.* Windsor: NFER-Nelson.

Elliott, C. D., Smith, P., & McCulloch, K. (1996). *British ability scales II.* Windsor, UK: NFER-Nelson.

Ellis, N. (1990). Reading, phonological skills and short-term memory: Interactive tributaries of development. *Journal of Research in Reading, 13*, 107–122.

Fidler, D. J., Most, D. E., & Guiberson, M. M. (2005). Neuropsychological correlates of word identification in Down syndrome. *Research in Developmental Disabilities, 26*, 486–501.

Fowler, A. E. (1990). Language abilities in children with Down syndrome: Evidence for a specific syntactic delay. In D. Cicchetti & M. Beeghly (Eds.), *Children with Down syndrome: A developmental perspective* (pp. 302–328). Cambridge, UK: Cambridge University Press.

Fowler, A. (1991). How early phonological development might set the stage for phoneme awareness. In S. Brady & D. Shankweiler (Eds.), *Phonological processes in literacy* (pp. 97–118). Hillsdale, NJ: Lawrence Erlbaum.

Fowler, A., Doherty, B., & Boynton, L. (1995). The basis of reading skill in young adults with Down syndrome. In L. Nadel & D. Rosenthal (Eds.), *Down syndrome: Living and Learning* (pp. 182–196). New York: Wiley-Liss.

Frith, U. (1985). Beneath the surface of developmental dyslexia. In K. E. Patterson, J. C. Marshall & M. Coltheart (Eds.), *Surface dyslexia* (pp. 301–330). Hove, UK: Lawrence Erlbaum Associates.

Gathercole, S. (2006). Nonword repetition and word learning: The nature of the relationship. *Applied Psycholinguistics, 27*, 513–543.

Gathercole, S. E., & Baddeley, A. D. (1994). Phonological working memory: A critical building block for reading and vocabulary acquisition? *European Journal of Psychology of Education, 8*, 259–272.

Glenn, S., & Cunningham, C. (2005). Performance of young people with Down syndrome on the Leiter-R and British picture vocabulary scales. *Journal of Intellectual Disability Research, 49*, 239–244.

Goetz, K., Hulme, C., Brigstocke, S., Carroll, J. M., Nasir, L., & Snowling, M. (2008). Training reading and phoneme awareness skills in children with Down syndrome. *Reading and Writing: An Interdisciplinary Journal, 21*, 395–412.

Goldman, R., & Fristoe, M. (2000). *Goldman-Fristoe test of articulation* (2nd ed.). Circle Pines, MN: American Guidance Service.

Groen, M. A., Alku, P., & Bishop, D. V. M. (2008). Lateralisation of auditory processing in Down syndrome: A study of T-complex peaks Ta and Tb. *Biological Psychology, 79*, 148–157.

Groen, M., Laws, G., Nation, K., & Bishop, D. V. M. (2006). A case of exceptional reading accuracy in a child with Down syndrome — Underlying skills and the relation to reading comprehension. *Cognitive Neuropsychology, 23*, 1190–1214.

Gunn, P., & Crombie, M. (1996). Language and speech. In B. Stratford & P. Gunn (Eds.), *New approaches to Down syndrome* (pp. 249–267). Cassell: London.

Jarrold, C., & Baddeley, A. D. (1997). Short-term memory for verbal and visuo-spatial information in Down's syndrome. *Cognitive Neuropsychiatry, 2*, 101–122.

Jarrold, C., Purser, H. R. M., & Brock, J. (2006). Short-term memory in Down syndrome. In T. P. Alloway & S. E. Gathercole (Eds.), *Working memory and neurodevelopmental conditions* (pp. 239–266). Hove, East Sussex: Psychology Press.

Jernigan, T. L., Bellugi, U., Sowell, E., Doherty, S., & Hesselink, J. R. (1993). Cerebral morphological distinctions between Williams and Down syndrome. *Archives of Neurology, 50*, 186–191.

Jiang, Z. D., Wu, Y. Y., & Liu, X. Y. (1990). Early development of brainstem auditory evoked potentials in Down's syndrome. *Early Human Development, 23*, 41–51.

Kay-Raining Bird, E., Cleave, P., & McConnell, L. (2000). Reading and phonological awareness in children with Down syndrome: A longitudinal study. *American Journal of Speech-Language Pathology, 9*, 319–330.

Kay-Raining Bird, E., Cleave, P., White, D., Pike, H., & Helmkay, A. (2008). Written and oral narratives of children and adolescents with Down syndrome. *Journal of Speech, Language, and Hearing Research, 51*, 436–450.

Kennedy, E. J., & Flynn, M. C. (2003). Training phonological awareness skills in children with Down syndrome. *Research in Developmental Disabilities, 24*, 44–57.

Kittler, P. M., Ho, T. T. P., Gardner, J. M., Miroshnichenko, I., Gordon, A., & Karmel, B. Z. (2009). Auditory brainstem evoked responses in newborns with Down syndrome. *American Journal on Intellectual and Developmental Disabilities, 114*, 393–400.

Laws, G. (2004). Contributions of phonological memory, language comprehension and hearing to the expressive language of adolescents and young adults with Down syndrome. *Journal of Child Psychology and Psychiatry, 45*, 1085–1095.

Laws, G., & Bishop, D. V. M. (2003). A comparison of language abilities in adolescents with Down syndrome and children with specific language impairment. *Journal of Speech, Language, and Hearing Research, 46*, 1324–1339.

Laws, G., & Bishop, D. V. M. (2004). Verbal deficits in specific language impairment and Down syndrome: A comparative review. *International Journal of Language & Communication Disorders, 39*, 423–451.

Laws, G., Buckley, S. J., MacDonald, J., & Broadley, I. (1995). The influence of reading instruction on language and memory development in children with Down syndrome. *Down Syndrome Research and Practice, 3*, 59–64.

Laws, G., Byrne, A., & Buckley, S. (2000). Language and memory development in children with Down syndrome at mainstream and special schools: A comparison. *Educational Psychology, 20*, 447–457.

Laws, G., Fisher, A., Guillaume, S., Lombard, F., & Nye, J. (2010). *A longitudinal study of oral language, short term memory and reading development in children with Down syndrome and children with specific language impairment*. Manuscript in preparation.

Laws, G., & Gunn, D. (2002). Relationships between reading, phonological skills and language development in individuals with Down syndrome: A five year follow up study. *Reading and Writing, 15*, 527–548.

Laws, G., & Gunn, D. (2004). Phonological memory as a predictor of language comprehension in Down syndrome: A five year follow up study. *Journal of Child Psychology and Psychiatry, 45*, 326–337.

Lemons, C. J., & Fuchs, D. (2010). Phonological awareness of children with Down syndrome: Its role in learning to read and the effectiveness of related interventions. *Research in Developmental Disabilities, 31*, 316–330.

Martin, G. E., Klusek, J., Estigarribia, B., & Roberts, J. E. (2009). Language characteristics of individuals with Down syndrome. *Topics in Language Disorders, 29*, 112–132.

Metsala, J. L. (1999). Young children's phonological awareness and nonword repetition as a function of vocabulary development. *Journal of Educational Psychology, 91*, 3–19.

Miller, J. F. (1999). Profiles of language development in children with Down syndrome. In J. F. Miller, M. Leddy & L. A. Leavitt (Eds.), *Improving the communication of people with Down syndrome* (pp. 11–39). Baltimore: Paul Brookes Publishing.

Morris, J. K., & Alberman, E. (2009). Trends in Down's syndrome live births and antenatal diagnoses in England and Wales from 1989 to 2008: Analysis of data from the National Down Syndrome Cytogenetic Register. *British Medical Journal, 339*, b3794.

Nadel, L. (2003). Down's syndrome: A genetic disorder in biobehavioral perspective. *Genes, Brain, and Behavior, 2*, 156–166.

Nagy, W. E., Herman, P. A., & Anderson, R. C. (1985). Learning words from context. *Reading Research Quarterly, 20*, 233–253.

Oelwein, P. L. (1995). *Teaching reading to children with Down syndrome: A guide for parents and teachers*. Bethesda, MD: Woodbine House.

Papavassiliou, P., York, T. P., Gursoy, N., Hill, G., Nicely, L. V., Sundaram, U., et al. (2009). The phenotype of persons having mosaicism for trisomy 21/Down syndrome reflects the percentage of trisomic cells present in different tissues. *American Journal of Medical Genetics. Part A, 149A*, 573–583.

Patterson, D. (2009). Molecular genetic analysis of Down syndrome. *Human Genetics, 126*, 195–214.

Pennington, B. F., Moon, J., Edgin, J., Stedron, J., & Nadel, L. (2003). The neuropsychology of Down syndrome: Evidence for hippocampal dysfunction. *Child Development, 74*, 75–93.

Perfetti, C. A., Beck, I., Bell, L., & Hughes, C. (1987). Phonemic knowledge and learning to read are reciprocal: A longitudinal study of first grade children. *Merrill-Palmer Quarterly, 33*, 283–319.

Pinter, J. D., Eliez, S., Schmitt, J. E., Capone, G. T., & Reiss, A. L. (2001). Neuroanatomy of Down's syndrome: A high-resolution MRI study. *The American Journal of Psychiatry, 158*, 1659–1665.

Reeves, R. H., & Garner, C. C. (2007). A year of unprecedented progress in Down syndrome basic research. *Mental Retardation and Developmental Disabilities Research Reviews, 13*, 215–220.

Roberts, J. E., Price, J., & Malkin, C. (2007). Language and communication development in Down syndrome. *Mental Retardation and Developmental Disabilities, 13*, 26–35.

Roch, M., & Jarrold, C. (2008). A comparison between word and nonword reading in Down syndrome: The role of phonological awareness. *Journal of Communication Disorders, 41*, 305–318.

Roid, G. M., Miller, L. J., & Lucy, J. (1997). *Leiter international performance scale-revised.* Wood Dale, ILL: Stoelting Co.

Roizen, N. (1997). Hearing loss in children with Down syndrome: A review. *Down Syndrome Quarterly, 2*, 1–4.

Rondal, J. A. (1995). *Exceptional language development in Down syndrome: Implications for the cognition–language relationship.* Cambridge: Cambridge University Press.

Rosin, M., Swift, E., Bless, D., & Vetter, D. (1988). Communication profiles in adolescents with Down syndrome. *Journal of Childhood Communication Disorders, 12*, 49–64.

Sloper, P., Cunningham, C., Turner, S., & Knussen, C. (1990). Factors related to the academic attainments of children with Down's syndrome. *The British Journal of Educational Psychology, 60*, 284–298.

Snowling, M., Bishop, D. V. M., & Stothard, S. E. (2000). Is pre-school language impairment a risk factor for dyslexia in adolescence? *Journal of Child Psychology and Psychiatry, 41*, 587–600.

Snowling, M. J., Chiat, S., & Hulme, C. (1991). Words, nonwords and phonological processes: Some comments on Gathercole, Willis, Emslie, & Baddeley. *Applied Psycholinguistics, 12*, 369–377.

Snowling, M. J., & Hulme, C. (2005). Learning to read with a language impairment. In M. J. Snowling & C. Hulme (Eds.), *The science of reading: A handbook.* Oxford: Blackwell Publishing Ltd.

Snowling, M. J., Hulme, C., & Mercer, R. (2002). A deficit in rime awareness in children with Down syndrome. *Reading and Writing, 15*, 471–495.

Stanovich, K. E. (1986). Matthew effects in reading: Some consequences of individual differences in the acquisition of reading. *Reading Research Quarterly, 21*, 360–407.

Tager-Flusberg, H. (1999). Language development in atypical children. In M. Barrett (Ed.), *The development of language* (pp. 311–348). London: UCL Press.

van Borsel, J. (1996). Articulation in Down's syndrome adolescents and adults. *European Journal of Disorders of Communication, 31*, 414–444.

van Bysterveldt, A. K., Gillon, G. T., & Moran, C. (2006). Enhancing phonological awareness and letter knowledge in pre-school children with Down syndrome. *International Journal of Disability, Development and Education, 53*, 301–329.

Vicari, S., Caselli, M. C., & Tonucci, F. (2000). Asynchrony of lexical and morphosyntactic development in children with Down syndrome. *Neuropsychologia, 38*, 634–644.

Whitehouse, A. J. O., Line, E. A., Watt, H. J., & Bishop, D. V. M. (2009). Qualitative aspects of developmental language impairment relate to language and literacy outcome in adulthood. *International Journal of Language & Communication Disorders, 44*, 489–510.

Ypsilanti, A., & Grouios, G. (2008). Linguistic profile of individuals with Down syndrome: Comparing the linguistic performance of three developmental disorders. *Child Neuropsychology, 14*, 148–170.

WILLIAMS SYNDROME

Deborah M. Riby* and Melanie A. Porter[†]

* SCHOOL OF PSYCHOLOGY, NEWCASTLE UNIVERSITY, UNITED KINGDOM
[†] DEPARTMENT OF PSYCHOLOGY, MACQUARIE UNIVERSITY, AUSTRALIA

I. Introduction

Williams syndrome (WS) is a neurodevelopmental disorder that was first identified in the 1960s (Williams, Barrat-Boyes, & Lowe, 1961). This genetic disorder is caused by the deletion of approximately 25 genes on

Advances in Child Development and Behavior
Patricia Bauer : Editor

chromosome 7 (7q11.23; Donnai & Karmiloff-Smith, 2000) and has a reported prevalence of 1 in 20,000, occurring sporadically within the general population (Morris, Demsey, Leonard, Dilts, & Blackburn, 1988). The genetic etiology of the disorder was unknown until the early 1990s when researchers discovered that individuals with WS invariably had a deletion of the elastin (ELN) gene on the long arm of chromosome 7 (Ewart *et al.*, 1994). By using genetic (fluorescent *in situ* hybridization, FISH) testing, it is now known that about 97% of individuals with WS have one copy of the ELN gene missing (see Lowery *et al.*, 1995). Before the availability of FISH testing, the disorder was identified as a condition involving a major heart defect (supravalvular aortic stenosis, SVAS; narrowing of the arteries that constricts blood flow), intellectual difficulty, and a distinct facial dysmorphology (see Bellugi, Klima, & Wang, 1996). For example, children with the disorder have been referred to as having an "elfin" face (e.g., Lenhoff, Perales, & Hickok, 2001); the characteristics of which may include a broad forehead, prominent eyes, flat nasal bridge, full cheeks, and a wide mouth (Morris & Mervis, 1999, and also see Hammond *et al.*, 2005). Medically, the most serious health problems are often related to the presence of the previously mentioned SVAS that may require surgery. Critically, however, WS is a multisystem disorder; for example, the severe cardiac abnormalities already mentioned occur alongside musculoskeletal problems (e.g., hyperextensive joints, small stature, delayed growth), gastrointestinal illnesses (e.g., hypercalcaemia, feeding problems in infancy), severe problems with sleep, problems with vision (e.g., strabismus, hyperopia), and issues with auditory stimuli processing (e.g., hyperacusis; Levitin, Cole, Lincoln, & Bellugi, 2005). For this range of problems, individuals with the disorder should be monitored regularly (see Morris & Mervis, 1999).

It was not until the middle to the end of the 1980s that researchers made considerable advances in understanding some of the neuropsychological features associated with WS. Since this time, there has been great progress in unearthing some of the major components of the WS cognitive and behavioral phenotypes and possible underlying neural mechanisms. In fact, WS leads the way in receiving attention and scrutiny from cognitive scientists and neuropsychologists (for a comprehensive review see Martens, Wilson, & Reutens, 2008). Those who have been interested in WS have used the disorder as a model for both typical and atypical behavior and development.[1] In this chapter, we explore the current state of knowledge regarding the cognitive, social, and behavioral profiles that are associated with WS and research based interventions and clinical

---

[1]Including arguments for or against modularity of the mind, see Temple and Clahsen (2002) and Pinker (1999), and for a discussion of these issues refer to Brock (2007).

approaches with this group. We take in turn several phenotypes (e.g., cognitive, social, behavioral) that have been linked to the disorder, first describing the range of features associated with that phenotype and then considering intervention implications related to each of these areas at the end of the chapter.

II. Cognitive Profile

A. IQ

Intellectually, individuals with WS tend to function at the level of a mild to moderate disability with Full Scale IQ (FSIQ) reported to be within the range of 50–60. However, individual variability will mean that FSIQ ranges between 20 (severe to profound intellectual disability) to 100 (average intellect) (see Bellugi, Mills, Jernigan, Hickok, & Galaburda, 1999; Porter & Dodd, in press; Searcy *et al.*, 2004), although such extremes are relatively rare. Having systematically reviewed studies exploring IQ in WS, Martens and colleagues report a mean FSIQ of 55 across 46 published articles (IQ ranging between 42 and 68). Research has indicated that this level of intellectual capability is relatively stable across the lifespan (e.g., see research with older individuals, Howlin, Davies, & Udwin, 1998; and see longitudinal research on children and adults, Porter & Dodd, in press). As will become clear in the following sections of this chapter, it should be noted that FSIQ is perhaps not the most meaningful ability measure for individuals with the syndrome.

A feature of WS cognition that has captured researcher's attention has been the relative difference between capabilities of language and visuospatial cognition. Going back to the systematic review compiled by Martens *et al.* (2008), across a large range of studies the average reported verbal intelligence quotient (VIQ) was 63 (ranging from 45 to 109), while the average reported performance intelligence quotient (PIQ) was lower at 55 (ranging 41–75). As these levels indicate, it is critical to keep in mind that any difference between functioning in these two domains is not simply a question of impaired versus spared functions. Indeed, the disparity between functioning in language and visuospatial domains occurs in comparison with the general level of mild–moderate intellectual disability that individuals present with. So the characteristic pattern of WS cognition is exemplified by greater proficiency within the language (auditory–verbal) than the visuospatial domain, at least on average (Bellugi, Bihrle, Jernigan, Trauner, & Doherty, 1990; Mervis & Klein-Tasman, 2000). A clear example of the dissociation was given by Bellugi *et al.* (1999) when they required individuals with WS to verbally describe and draw an elephant. The verbal description was lengthy, but the drawing revealed poor linkage between the elephant's component parts. It could be proposed that the dissociation between these capabilities is driven by severe deficits of visuo-spatial cognition and fine-motor deficits, as opposed to any evidence of remarkably preserved language skill. Indeed, further inspection of the verbal description suggested skills far below chronological age expectations.

We must bear in mind when seeing reported intelligence quotients for groups of individuals with WS, that a single mean IQ level will not truly represent the variability that will exist within the cohort (in much the same way as the heterogeneity between individuals with other disorders of development). Indeed we should note that some individuals with WS will display similar verbal and nonverbal (spatial) capabilities, and in rare instances, some individuals may show better nonverbal than verbal capabilities (see Howlin *et al.*, 1998; Porter & Coltheart, 2005; Porter & Dodd, in press). Importantly though, the WS cognitive profile, characterized by a dissociation of functioning in language and visuospatial domains, is widely reported across a large number of studies and has provided the impetus for in-depth evaluations of the WS cognitive profile.

B. LANGUAGE

So how good are the language skills of individuals with WS? Is language really "spared" or do deficits exist within this domain? We have already noted that it is important to consider language functioning in relation to general cognitive abilities and it should also be considered relative to the individuals' chronological age. However, many preliminary studies of language skill associated with WS used descriptive terms such as "spared" or "intact" (e.g., Bellugi, Lai, & Wang, 1997; Bellugi, Marks, Bihrle, & Sabo, 1988). Importantly though, there are likely to be relative strengths and weaknesses of language skill even within this one domain of functioning. It should also be considered that relative strengths for some language skills may (i) influence processing in other domains of cognition and (ii) mask more subtle but significant language deficits. For example, a relative strength of receptive vocabulary is widely reported (Bellugi, Marks, *et al.*, Brock, 2007; Porter & Coltheart, 2005) and this ability, along with a friendly and sociable demeanor (outlined later in this chapter), a strong interest in conversing with others, excessive talkativeness, generally clear articulation skills, and speech fluency (e.g., see Udwin, Yule, & Martin, 1987), may make individuals with WS appear more linguistically able than they really are. As for many areas of cognition associated with WS, research has recently used assessments that can sensitively assess different components of language skill to delve deeper into those skills originally considered "spared."

In this section, we focus on evidence from different components of auditory–verbal skill, including important precursors of language development, before evaluating the possible use of language intervention strategies.

1. Precursors of Language Skill and Early Language Development

We first touch on two early cognitive language precursors, namely auditory processing and phonological short-term memory. Both these skills are considered important for the development of early language abilities in typical development (e.g., Gathercole & Baddeley, 1989; Mehler *et al.*, 1988). How do individuals with WS fare with regards to these two skills? Individuals with the disorder are reported to show a relative strength for auditory processing, such as rhythm and the perception of tone and pitch (Don, Schellenberg, & Rourke, 1999; Lenhoff *et al.*, 2001; Levitin & Bellugi, 1998; Porter & Coltheart, 2005). There are also reports that they are able to apply emotional prosody to their communication, indicating that they may be able to apply rhythm and auditory cues in their own speech (Reilly, Harrison, & Klima, 1995). Therefore, later abilities of auditory processing may indicate that these skills have played a role in the early development of language skills; however, further work is required to explore the role of these specific skills at critical time points of language development.

Similarly, individuals with WS are said to display a relative strength in their phonological short-term memory ability; at least compared to their other intellectual abilities (Porter & Coltheart, 2005; Porter & Dodd, in press), their visuospatial short-term memory (e.g., Jarrold, Baddeley, & Hewes, 1999; Wang & Bellugi, 1994) and compared to individuals with Down syndrome (DS) (Jarrold *et al.*, 1999; Wang & Bellugi, 1994).[2] Importantly though, a relative strength of phonological short-term memory certainly does not occur for all individuals with WS and some researchers have failed to find evidence of such strength (see Brock, 2007 for a review). Therefore, individual variation will occur and these

---

[2]It should be noted, as Brock (2007) points out, that phonological short-term memory is a known weakness in Down syndrome (DS) relative to IQ and therefore, although making comparisons between this skill in WS and DS allows us to question the specificity of phonological short-term memory skill to WS, using this comparison group tells us little about the typicality ability.

individual differences may influence the course of language development
and lead to variable outcomes of language capability.

It is important to note that despite these two precursors of language
development being considered relative strengths for individuals with WS
(especially later in life), the onset of language and the achievement of
early language milestones are almost always delayed (Mervis & Klein-
Tasman, 2000, Singer-Harris, Bellugi, Bates, Jones, & Rossen, 1997).
Given the possibility of additional delay in the development of early nonverbal social communicative tools such as pointing gestures and joint
attention (Laing *et al.*, 2002, Rowe, Peregrine, & Mervis, 2005) and
similarities of early communication atypicalities between young children
with WS and Autism (e.g., Klein-Tasman, Mervis, Lord, & Phillips,
2007), this may be a particularly vulnerable time for young children with
WS. At this point, a high proportion of children with WS may require
early speech therapy interventions (we return to the issue of language
intervention later in this chapter).

So how about other (more sophisticated) language skills once an individual has mastered early language capabilities? We take a range of language skills in turn to illustrate language capabilities as well as those
components that might prove more problematic. Each of these issues
may be relevant to the design of interventions targeting language in WS.

2. Vocabulary

One well established finding is that individuals with WS perform well on
tests of naming and receptive vocabulary, such as the Boston Naming
Test, the Peabody Picture Vocabulary Test, and the British Picture
Vocabulary Scale, often performing at or close to chronological age levels
(Bellugi *et al.*, 1990; Mervis, Robinson, Bertrand, & Morris, 2000; Temple,
Almazan, & Sherwood, 2002). Moreover, this relative strength of receptive
vocabulary tends to persist with age (Brock, Jarrold, Farran, Laws, &
Riby, 2007; Porter & Dodd, in press). Contrary to this relative strength
shown on standardized assessments, some parents report receptive language difficulties in their children using the Vineland Adaptive Behavior
Scale (Howlin *et al.*, 1998). This insight may relate to the difference
between direct testing versus the impressions of parents who may find it
hard to disentangle the child's vocabulary skills from other aspects of their
everyday behavior.

Individuals with WS display poorer expressive than receptive vocabulary. Expressive vocabulary skills are, however, more advanced than those
of children with DS who are of the same chronological age (e.g., Mervis &
Robinson, 2000; Singer-Harris *et al.*, 1997). Importantly though, individual

variation will exist for both receptive and expressive vocabulary abilities (e.g., Mervis & Robinson, 2000; Porter & Coltheart, 2005) and these individual differences should not be ignored when considering the needs of individuals with the disorder.

3. Grammar and Syntax

Early development suggests consistency between vocabulary development (considered above) and grammatical complexity, with both syntax and morphology generally consistent with mental age level (see Brock, 2007, for a review). However, some difficulties with comprehension and expression of more grammatically complex sentences have been noted, at least in subsets of individuals (Joffe & Varlokosta, 2007; Mervis & Becerra, 2007). For example, Frank (1983); as cited by Semel and Rosner (2003) observed that approximately one-fifth of children with WS (aged 4 and 17 years) displayed impairments in complex syntactical development. Moreover, while individuals with WS tend to outperform those with DS on tests of receptive grammar (e.g., see Bellugi *et al.*, 1990), they tend to perform well below chronological age expectations and display distinct error patterns (e.g., difficulties with relative clauses and embedded sentences; Karmiloff-Smith *et al.*, 1997; Mervis, Morris, Bertrand, & Robinson, 1999). A tendency to display problems in the grammatical use of morphemes, incorrect tense markings, and incorrect usage of personal pronouns is also reported (see Semel & Rosner, 2003 for a review) and suggests subtle problems with a range of sophisticated components of language.

4. Semantics

Early studies indicated that individuals with WS performed relatively well on semantic fluency tasks compared to their general level of intelligence and were able to name a large number of category exemplars within a given time period (e.g., naming as many animals as possible in 60 s). However, investigation of the category items named by individuals with the disorder has suggested the possible involvement of unusual category exemplars (e.g., Reilly, Klima, & Bellugi, 1990; Temple *et al.*, 2002). Along with reports of an unusual choice of words in spontaneous conversation (e.g., Bellugi, Wang, & Jernigan, 1994) this has been used to propose that the processing of semantic information occurs atypically. For example, producing more unusual words may indicate an atypically reduced sensitivity to word frequency. However, not all studies have found evidence of semantic anomalies or rare word use on semantic fluency tasks (e.g., Jarrold, Hartley, Phillips, & Baddeley, 2000; see Mervis & Becerra, 2007 for a review) or in elicited narratives (Stojanovik & van Ewijk, 2008). It is worth noting once again, that individual variability within WS (especially in terms of general language skill) will be central to the discrepancy between reported studies. For those individuals who do tend to use relatively rare or unusual words, speech and language therapy may be a useful intervention relatively early in language development. We note later in this chapter how possible atypicalities of semantics may play a role in reading development and, therefore, how this skill may be an important component of literacy interventions.

5. Pragmatics

In addition to abnormalities of language content, impairments in the social *use* of language (pragmatics) are evident. Strikingly, Laws and Bishop (2004) found that 79% of children and young adults with WS met the criteria for a pragmatic language impairment based on parent and teacher ratings on the Children's Communication Checklist (CCC; Bishop & Baird, 2001). This level was compared to 50% for DS and 41% of individuals with Specific Language Impairment. Using the revised version of the Children's Communication Checklist (the CCC-2), Philofsky, Fidler and Hepburn (2007) also reported pragmatic language impairments, albeit not as severe as those for individuals who were functioning on the autistic spectrum. These parental insights suggest that social language use of individuals with WS may be far from typical and, indeed, these peculiarities are deeply entwined with atypicalities of social engagement style reported later in this chapter.

Reilly, Losh, Bellugi and Wulfeck (2004), as well as Stojanovik, Perkins and Howard (2004) went beyond parental report measures to experimentally investigate pragmatic language skills. Both research groups reported disorganized language content and poorly structured discourse in the narratives produced by children with WS, even when compared to children with Specific Language Impairment. Stojanovik (2006) also utilized a more ecological setting, where they analyzed the conversational structure of five school-aged children with WS and reported that their conversations were less mature in structure and more inappropriate in language content compared to typically developing children and children with language impairments. Children with WS tended to provide too little information for their conversational partner, tended to over-rely on the conversational partner's lead and contribution, and tended to demonstrate overly literal misinterpretations. Therefore, subtle atypicalities of pragmatics may hinder the flow of social communication and have an impact upon shaping social interactions for individuals with the disorder.

6. Verbal Memory

While phonological short-term memory (or immediate memory span for auditory–verbal information) is generally reported to be at or above the level predicted by mental age (Porter & Coltheart, 2005), verbal working memory (being able to hold and manipulate auditory–verbal information in mind) is generally at, or below, mental age level (see Mervis, 2009 for a review). The difference between these abilities is likely to be the presence of the additional executive (manipulation) component in the latter of these tasks. We will return to problems with the executive control of attention and executive functioning later in this chapter (but see Rhodes, Riby, Park, Fraser, & Campbell, 2010). Executive skills, as well as difficulty combining verbal cues with other sources of information, may also contribute to the relative weakness of visual–auditory or cross-modal learning and delayed memory (Porter & Coltheart, 2005) in WS and may require intervention targeted toward these skills.

7. Typical versus Atypical Language Development

Having considered various components of language skill, we must take a moment to consider the critical role of developmental change. A longstanding debate within the WS literature has been whether we should consider components of language as merely impaired or intact (in relation to general intellect), or whether we should focus on the typical/atypical developmental pathway of language skill in this population. For example, how is language development shaped in WS?[3] Karmiloff-Smith, Brown, Grice and Paterson (2003) emphasize that because the beginning state of the WS brain is different to that of a typically developing brain, the processes by which language is acquired in WS are also highly likely to be different. Therefore, even in domains where individuals with WS show relative proficiency, it is likely that different developmental processes have shaped this outcome. As an example, studies have suggested that phonological processing abilities develop atypically in WS; including a reduced influence of long-term phonological and semantic knowledge on phonological short-term memory (Brock, 2007). Others also suggest that individuals with WS rely too heavily on their relatively

[3]For in-depth arguments see Karmiloff-Smith (1998); Karmiloff-Smith, Scerif, and Thomas (2002) and Thomas, Karaminis, and Knowland (2010).

good phonological processing and verbal memory skills during language acquisition, at the expense of relying on more conceptual aspects of cognition (Mervis, Robinson, Rowe, Becerra, & Klein-Tasman, 2003). Finally, it has been suggested that the development of lexical–semantic processing deviates from a "typical" path of development (Rossen, Klima, Bellugi, Bihrle, & Jones, 1996; Temple *et al.*, 2002). These developmental aspects are critical for the establishment of targeted intervention programs aimed at improving language capabilities. In designing such intervention programs or techniques we must be aware that milestones of language skill may be different when language is shaped by WS rather than typical development.

In summary, while there are some areas of relative strength within the domains of language and auditory verbal skills associated with WS, there are also quite significant impairments and language as a whole is likely not to follow a typical developmental pathway.

C. VISUOSPATIAL COGNITION

1. Global and Local Processing

We have reported that compared to verbal IQ (and particularly compared to receptive vocabulary), visuospatial cognition is, on average, a relative weakness for individuals with WS. Tasks that involve fine-motor skills or a construction element to successful task completion are especially difficult. For example, deficits have been reported in fine-motor skills, orientation discrimination, visuomotor integration, and spatial memory (Bellugi, Sabo, & Vaid, 1988; Farran, 2005; Farran & Jarrold, 2003; Farran, Jarrold, & Gathercole, 2003; Hoffman, Landau, & Pagani, 2003; Vicari, Bellucci, & Carlesimo, 2005). Deficits have also been noted in scanning and tracking of complex visual spatial information, with anecdotal reports suggesting that WS individuals often lose their place on a complex worksheet or a page of written text, or experience difficulty focusing in a busy environment. Visual scanning deficits have also been noted more formally on medical examination, at least in the form of saccadic eye movements (van der Geest, Lagers-van Haselen, & Frens, 2006).

Much work assessing visuospatial cognition in WS has used the distinction between global and local processing capabilities. For example, tasks may involve copying a stimuli image that contains both local and global elements. A traditional example of this is the Navon stimuli whereby small local elements link together to form a larger element (e.g., a large "N" shape made up of small "h" letters showing the distinction between the global shape, N, and the local elements, h). Although in typical development older children and adults process information at the global level faster than at the local level, individuals with WS do not show this "global precedence" (see Farran *et al.*, 2003). Instead, individuals with WS show a bias toward processing local features (Bihrle, Bellugi, Delis, & Marks, 1989). On Navon tasks that tap the distinction between local and global processing capacity, research has indicated that global deficits are indicative of delayed, rather than deviant, abilities as performance patterns tend to mirror those of younger typically developing children (Bertrand, Mervis, & Eisenberg, 1997; Georgopoulos, Georgopoulos, Kuz, & Landau, 2004).

It has further been suggested that this type of global processing deficit is more likely for tasks that involve a construction or replication element (e.g., drawing or Block Design tasks), than perceptual tasks; perhaps due to problems of (i) fine-motor skill, (ii) planning motor responses (Hoffman *et al.*, 2003), or (iii) problems alternating between the global and local elements (Pani, Mervis, & Robinson, 1999; Porter & Coltheart, 2006). In a review of the literature, Farran and Jarrold (2003) suggest that although adults with WS have difficulty in drawing or replicating global aspects of a stimulus, they are able to detect and thus perceptually discriminate both global and local features. This was also supported by Porter and Coltheart (2006), where individuals with WS were able to draw both global and local features when asked to copy Navon stimuli, but they often drew them separately without integrating the components. A clear example of this is provided in Figure 1, showing drawings completed by six individuals with WS in their attempt to draw a rectangle made up of small crosses or a large square made up of small squares. Porter and Coltheart (2006) suggested that a bias toward processing local components is "attention" based. They compared and contrasted global and local attention, perception, and construction in individuals with WS, DS, or Autism and found an attention bias toward local features in both WS and Autism and a global attention bias in DS. Interestingly the attention bias did not extend to perception. Integration deficits were seen for a subset of the participants with WS, but there were dissociations between the integration of information for perception and construction. Thus, it is well established that individuals with WS display a range of nonverbal deficits including fine-motor difficulties and spatial integration deficits.

2. Nonverbal Memory

In contrast to phonological short-term memory detailed earlier, immediate memory span for spatial information appears to be relatively impaired in WS (Jarrold *et al.*, 1999; Rhodes *et al.*, 2010; Wang & Bellugi, 1994; Vicari, Bellucci, & Carlesimo, 2006a; Vicari, Bellucci, & Carlesimo,

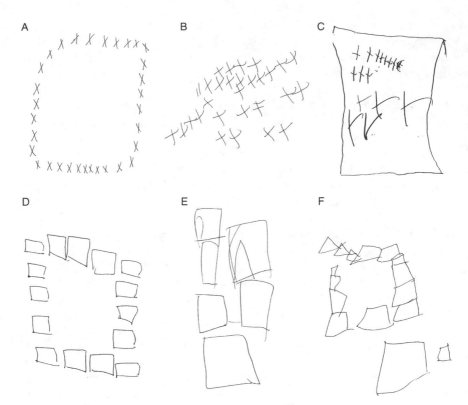

Fig. 1. Drawings by six individuals with Williams syndrome. Individuals were instructed to draw a rectangle made up of small crosses (A–C) or a large square made up of small squares (D–F).

2006b). These problems are further exacerbated if there is an executive (working memory) component to the spatial memory task (Rhodes *et al.*, 2010). However, short-term memory for visual objects is at or even slightly above mental age levels (Porter & Coltheart, 2005; Porter & Dodd, in press). It is most likely that poor performance on visuospatial memory tasks is derived from deficits processing the spatial information conveyed in the task. Indeed when required to match target spatial arrangements individuals with WS perform poorly in a 0-s delay task and even more poorly when an additional memory component is introduced (e.g., a 4-s delay between stimulus and target, Rhodes *et al.*, 2010). Thus, problems making judgments of a spatial nature impact upon other cognitive capabilities such as spatial working memory, and, as we shall see below, face perception.

3. Face Perception

In contrast to the visuospatial weaknesses already reported, face perception is suggested to be a selective strength within the visuospatial domain (also linking to the social phenotype of the disorder reported later in this chapter). During the 1990s, initial exploration of the cognitive profile associated with WS suggested an "island of sparing" with relation to face recognition. Both children and adults with WS perform relatively well on standard face recognition measures such as the Benton Facial Recognition Test (Benton, Hamsher, Varney, & Spreen, 1983) and the Warrington Recognition Memory Test (Warrington, 1984) compared to their typically developing peers and relative to individuals with other neurodevelopmental disorders. However, the studies reporting evidence of relatively "intact" face recognition skill did not reveal *how* individuals with WS encoded faces. Although task performance may sometimes appear to be within a "typical" range, this does not mean the face has been encoded using a typical strategy. It is now recognized that at least some aspects of face processing are developmentally delayed or abnormal (Karmiloff-Smith *et al.*, 2004; Porter, Shaw, & Marsh, 2010; Tager-Flusberg, Plesa-Skwerer, & Joseph, 2006 for recent discussions).

The main abnormality that has been proposed in terms of face perception in WS is the disruption of configural structural encoding (e.g., Deruelle, Mancini, Livet, Casse-Perrot, & de Schonen, 1999; Karmiloff-Smith *et al.*, 2004, Wang, Doherty, Rourke, & Bellugi, 1995). Configural processing requires assessment of spatial relations between facial features (e.g., eye to eye distance). Typical adults' process faces based on their configural properties (Young, Hellawell, & Hay, 1987) and for typically developing children configural processing develops with age and expertise (Carey, 1981). Between the ages of about 6 and 10 years individuals who are developing typically begin to use configural information in a systematic way (Mondloch, Le Grand, & Maurer, 2002). One way to explore configural processing is to invert a face as inversion disrupts the processing of configural processing (Leder & Bruce, 2000). The face "inversion effect" refers to a performance reduction when faces are inverted compared to when they are upright (Yin, 1969). There is inconsistency regarding the effect of inversion on face perception in WS; while some studies report a reduced or missing inversion effect (e.g., Deruelle *et al.*, 1999; Elgar & Campbell, 2001; Karmiloff-Smith, 1997) other studies reported a more "typical" effect (e.g., Riby, Doherty-Sneddon, & Bruce, 2009; Rose *et al.*, 2006). The differences across these studies may relate to both the stimuli used and the individual differences corresponding to participant age and ability. However, another way of assessing configural processing is to require individuals to detect subtle spacing changes made to faces and in this type of study both children with WS (Deruelle *et al.*, 1999) and adults with WS (Karmiloff-Smith *et al.*, 2004) show deficits of detecting configural changes.

In further assessment of the development of configural face processing skills in adults with WS, Karmiloff-Smith and colleagues (2004; $n = 12$, age range 16–51 years) found not only evidence of developmental delay but also an atypically reduced reliance upon configural processing. So it seems that individuals with WS have problems with processing the spatial relations within faces in much the same way that they have problems processing spatial aspects of other nonsocial objects as previously reported.

The final component of structural encoding to consider is the use of holistic processing. In this account, the face is recognized as a whole image and template-matching may be used as a rapid route to recognition (Tanaka & Farah, 1993). Holistic processing in WS was first assessed in adolescents and adults with WS by Tager-Flusberg, Plesa Skwerer, Faja and Joseph (2003; age range 12–36 years), who used a part–whole paradigm; participants learnt a set of target faces and associated names and then identified features of a target face presented either in isolation (e.g., Which is Fred's nose?) or in the context of a whole face (e.g., Which person is Fred?). The stimuli are often presented in upright and inverted conditions. Participants with WS tend to perform less accurately than participants of comparable chronological age in all conditions. Importantly though, the same results *pattern* was evident across groups. The authors noted that a whole face advantage for upright but not inverted images was indicative of holistic processing in both WS and typical development. Annaz, Karmiloff-Smith, Johnson and Thomas (2009) followed up this finding in a younger age group of children with the disorder and found that participants with WS (age range 5–12 years) showed a "typical" pattern of recognizing facial features in isolation better than in the context of a whole face and performed better for upright than inverted trials. The group showed no significant change of performance across the age range tested. The analyses suggested that individuals with WS showed a slower rate of development for holistic face perception skills but that development did not deviate from a "typical" pathway.

So, from initial suggestions of an "island of sparing" for face perception in WS we now have considerable evidence to suggest a range of atypicalities to the way that faces are processed and these atypicalities relate more widely to (i) general developmental delay associated with the disorder and (ii) atypicalities of spatial perception. Some of these atypicalities may also be shaped by abnormalities of orienting to faces throughout development; from infancy (e.g., Mervis, Robinson, *et al.*, 2003; Mervis *et al.*, 2003) into adolescence and adulthood (e.g., Riby &

Hancock, 2008). It may be that a propensity toward increased attention to faces can prove useful in the design of engaging materials for social skills training and we will considered this possibility later in the current chapter.

D. ACADEMIC SKILLS

One area that might be particularly accommodating of interventions is the education provision of children and young people with a diagnosis of WS. In this section, we focus on literacy and numeracy as key indicators of academic achievement and abilities. We note both strengths and weaknesses and how these skills might feature in targeted intervention programs designed specifically to the needs of children and young people with WS.

1. Literacy

The WS language profile (described above) has important implications for the development of literacy in children with WS. With a reported relative strength in phonological short-term memory and a common advantage of some aspects of verbal ability over nonverbal abilities, one might expect relatively good reading skills. Indeed, reading abilities appear to be correlated with verbal intellect, as is the case in the typically developing population (Howlin *et al.*, 1998). However, it is critical to consider that: (i) reading involves a visuospatial element; (ii) reading involves auditory–visual cross-modal (integrative) learning and (iii) reading, especially reading comprehension, relies heavily on semantics and verbal working memory; all areas of relative impairment in WS. Also, there is some evidence to suggest that nonverbal (spatial) skills correlate more highly with reading abilities in WS than verbal intellect (Howlin *et al.*, 1998), suggesting that deficits of spatial ability play an important role in the development or demise of reading skills.

Consistent with relative strengths in phonological short-term memory and verbal intellect, phonemic awareness skills and single word reading abilities, at least on average, appear to be at or above mental age (e.g., see Mervis, 2009 for a review). However, as Mervis (2009) points out, mental age equivalence scores are commonly reported within the literature, but other indices such as standard scores may be more reliable (whereby the mean is 100 and standard deviation is 15). Without the use of standard normative scores, it is difficult to gauge whether reading ability is commensurate with the level expected for IQ. Also, even when standardized scores are reported, they are usually directly compared to other standardized scores (e.g., an IQ standard score of 63 may be directly compared to a reading score of 63). We know, however, that there is no direct relationship between standardized scores on IQ tests and standardized scores on academic tests (i.e., a standard score of 70 on an IQ test does not equate to an expectation that reading ability should be 70 to be on par with IQ). Instead, psychometric tables should be used to predict reading ability based on IQ. To the best of our knowledge, no study has assessed whether reading ability is commensurate with FSIQ or VIQ in WS using psychometric tables. Thus, it remains unclear whether single word reading abilities are consistent with FSIQ, VIQ, or indeed PIQ.

So, are reading abilities delayed or deviant in WS? Laing (2002) suggested that single word reading was a relative strength, but critically, despite reasonably good abilities, individuals with WS may arrive at their level of single word reading by an atypical route. Thus, it is possible that some aspects of reading development are delayed and others deviant. For example, it has been suggested that (similar to other phonological processes), phonological awareness skills are delayed, but are unlikely to develop atypically. In contrast, the relationship between reading and semantic processing *may* be atypical in WS. Semantics play a role in reading development; at least within the typically developing population (e.g., see Mervis, 2009). Therefore, it is possible that the atypical develop-ment of semantics has implications for the development of reading skills. Indeed, individuals with WS seem to rely less on semantics than typically developing children when learning to read (Laing, Hulme, Grant, & Karmiloff-Smith, 2001) and for some individuals with the disorder there may be evidence of atypical semantic errors being made[4] (Temple, 2003).

In summary, the jury is still out regarding whether (i) reading and reading comprehension abilities are consistent with intellectual capabilities in WS or (ii) whether reading develops typically or atypically. Again, there is striking variability of reading skills amongst individuals with the disorder and to accommodate this variability it has been noted that "reading skills range from an inability to read at all to age-appropri-ate decoding and comprehension" (Mervis, 2009; p. 150).

2. Spelling

There are limited studies exploring spelling abilities of children with WS. Howlin *et al.* (1998) found that spelling was significantly below the level of single word reading in this group, suggesting further studies are

---

[4]see Temple (2003) for the discussion of a case of Deep Dyslexia in WS suggesting a theory of "sloppy lexical access" or "looser semantic specifications."

warranted. Future studies should specifically explore whether there is an overreliance on phonics at the expense of lexical representations. Studies also need to consider the impact of fine-motor and spatial difficulties; for example, anecdotally children with WS make many spatial errors when trying to write letters (e.g., "b" for "d" or "p" for "q"), and letter sequencing errors are also reasonably commonly reported.

3. Numeracy

Despite frequent anecdotal reports of poor mathematical skills, there are a limited number of studies exploring mathematical ability within the syndrome. Limited evidence suggests that some aspects of mathematical skill are much poorer than predicted by overall intellectual abilities (Bellugi, Marks, *et al.*, 1988; Bellugi, Sabo, & Vaid, 1988; Howlin *et al.*, 1998; Paterson, Girelli, Butterworth, & Karmiloff-Smith, 2006, Udwin, Davies, & Howlin, 1996). These deficits may relate to the visuospatial and the spatial construction deficits outlined above. Generally, mathematical skills (or quantitative reasoning) are a significant weakness within academic capabilities. Mirroring the pattern seen within other domains of cognition, some components of mathematics are relatively more impaired than others (see O'Hearn & Landau, 2007). For example, a selection of studies involving infants and toddlers suggest that numerical magnitude (number line) and approximate number estimates may be more impaired than number recognition, memory for mathematical facts (addition and multiplication), and verbal encoding of numbers at this early stage of development (Ansari *et al.*, 2003; Krajcsi, Lukács, Igács, Racsmány, & Pléh, 2009, O'Hearn & Landau, 2007, and see O'Hearn & Luna, 2009 for a review). Moreover, individuals with WS appear to experience particular difficulties discriminating spatial arrays of larger numbers as opposed to smaller numbers (Van Herwegen, Ansari, Xu, & Karmiloff-Smith, 2008), perhaps at least partially reflecting (i) the mental capacity of the task, (ii) spatial attention deficits, or (iii) problems with visual tracking. Importantly, when considering a range of numerical and mathematical skills, it is important to remember that these skills will be affected not only by lower-level cognitive deficits including information processing, attention, and spatial skills, but also by higher-level (executive) functions such as working memory, impulsivity, nonverbal reasoning, and problem-solving deficits. It is therefore difficult to separate deficits of mathematical ability from other cognitive abilities or deficits. However, problems with number skills can be widespread and have implications for everyday functioning; specifically, individuals experience great difficulty telling the time and, as adults, the vast majority are unable to manage their own finances (Davies, Udwin, & Howlin, 1998; Udwin, 1990).

E. ATTENTION AND EXECUTIVE FUNCTIONING

There is a relative scarcity of research on attention and executive functioning abilities of individuals with WS across all age groups (but see Rhodes *et al.*, 2010; Rhodes *et al.*, in press). Importantly, distractibility, inattention, inflexibility, and impulsivity are areas that represent major difficulties for education (and link to wider issues of social behavior) for many individuals with the disorder, especially during childhood. Indeed these problems link to all domains of cognition reported in this chapter so far and are likely to have implications across the lifespan (e.g., being especially problematic in an educational setting and needing to be considered in the design of academic interventions, plus creating problems with independence later in life).

Work with toddlers who have WS emphasizes that attention characteristics may be specific to WS as opposed to other disorders (such as Fragile X syndrome; Scerif, Cornish, Wilding, Driver, & Karmiloff-Smith, 2004). Toddlers with WS regularly confused distracters with targets. The authors suggested that the pattern of results represented a range of deficits of visual and attentional processing. These deficits on tasks tapping the executive control of attention are likely to extend to older children and adults (Atkinson, 2000).

Parent report ratings and structured clinical interviews suggest high rates of inattention and distractibility and poor behavior and emotion regulation in children and young people with WS (Dodd & Porter, 2009; Leyfer, Woodruff-Borden, Klein-Tasman, Fricke, & Mervis, 2006; Porter, Dodd, & Cairns, 2008; Rhodes *et al.*, in press). In a latter section of this chapter, we will note the role of attention problems in the maladaptive behaviors associated with the disorder. For example, children with WS have been reported by their parents to show as severe problems with inattention and hyperactivity (measured using the Conners rating scale) as children with Attention Deficit Hyperactivity Disorder (ADHD) (Rhodes *et al.*, in press). These behaviors have clinical implications for the management and education of individuals with the disorder.

Linking to the specific types of attention problems that individuals with WS may experience, there is evidence of problems with executive functioning, including behavior and emotion regulation difficulties (Einfeld, Tonge, & Florio, 1997), poor response inhibition (Porter, Coltheart, &

Langdon, 2007), and attention shifting (Rhodes *et al.*, 2010) throughout development. Other aspects of executive functioning, such as nonverbal reasoning and concept formation, appear to be relatively more commensurate with general mental age ability (Porter & Coltheart, 2005). However, longitudinal evidence suggests increased executive functioning deficits with age, paralleling the typical maturation of the frontal lobes and perhaps a widening gap between WS individuals and their peers in adulthood (Porter & Dodd, in press).

III. Social Behaviors

In this section, we focus on behaviors linked to the WS social phenotype. Much work over the last decade has focused on social behaviors associated with WS, having spent much of the preceding decade focused on the cognitive profile detailed previously. Many individuals with WS show an intriguing profile of social attributes. They tend (both as children and adults) to be exceptionally friendly, outrageously outgoing, and highly sensitive to the feelings of people they interact with. Yet it therefore seems contradictory and somewhat confusing that so many young people with the disorder have extreme problems forming lasting peer relations (Davies *et al.*, 1998). For a group that is characterized by descriptions of being "people-oriented", "friendly" "empathetic", and "affectionate" (Gosch & Pankau, 1997; Klein-Tasman & Mervis, 2003; Tager-Flusberg & Sullivan, 2000; Tomc, Williamson, & Pauli, 1990) difficulties forming and maintaining relationships may appear somewhat surprising. Similarly, for a group characterized by these social qualities an extremely high number of adults with the disorder (almost 73%) experience social isolation (Davies *et al.*, 1998) and suffer from social anxieties and social withdrawal (Mervis & Klein-Tasman, 2000; Porter *et al.*, 2008). Although the descriptive terms used above may appear beneficial to social functioning, these characteristics represent a range of atypicalities that, in combination, create a distinct profile of social engagement behaviors. For example, this friendliness and an orientation to engage in interactions with both familiar and unfamiliar people go far beyond that which is expected in typical development. Overfriendliness and oversensitivity, coupled with mild–moderate intellectual difficulties, and subtle problems with socially related behaviors can be a real problem for individuals with the disorder as well as their carers and families (see Jawaid *et al.*, submitted).

It is important to note that atypicalities of social behavior emphasized in this section are not isolated to children with WS and extend to adults.

Emphasizing the existence of social disinhibition in adults that mirrors that reported for children, Davies *et al.* (1998) reported that 94% of adults (sample size $n = 70$) were socially disinhibited. Equally, problems with communication, independence, and socialization are an issue during adulthood (assessed using the Vineland Adaptive Behavior Scale; Howlin *et al.*, 1998). Adults with WS have fewer close friendships than adults with other forms of intellectual difficulty (Dykens & Rosner, 1999). These problems are highly likely to be implicated in the high dependency of adults with the disorder and the small number of completely independent adults who can function without significant assistance in their daily lives. This may be a timely concern due to the increased number of individuals with WS who are reaching older age due to advances of medical treatment that can target SVAS, the main medical concern for those with WS. Intervention strategies may need to be directed toward aiding older individuals with the disorder and in this way social dependency may become a behavior targeted for intervention.

A. HYPERSOCIABILITY

In 2000 Jones and colleagues were the first to use the term "hypersociable" to refer to individuals with WS. This term was used to characterize a propensity toward increased desirability for engaging in conversation with both familiar and unfamiliar people. This term has since been readily applied to individuals with the disorder across the lifespan and is now used to summaries social interest and social approach behaviors (e.g., Doyle, Bellugi, Korenberg, & Graham, 2004; Frigerio *et al.*, 2006; Jones *et al.*, 2000; Riby & Hancock, 2008). Some researchers have gone as far as to say that individuals with WS have a "compulsion" toward social interactions (Frigerio *et al.*, 2006) and that there is "social fearlessness" in this group (Haas *et al.*, 2010). This increased social drive or hypersociable behavior is present across age and culture (e.g., Zitzer-Comfort, Doyle, Masataka, Korenberg, & Bellugi, 2007). It is useful to bear in mind that this is rather different from the social withdrawal shown by individuals who are functioning on the autistic spectrum (e.g., Frith, 1989), very different to the aversion to social engagement seen in Fragile X syndrome (e.g., Hall, DeBernardis, & Reiss, 2006), more exaggerated than the social behaviors shown by individuals with DS (e.g., Rosner, Hodapp, Fidler, Sagun, & Dykens, 2004) and also very different from the type of social behavior that we associate with typical development.

B. SOCIAL APPROACH

In WS, social approach behaviors have been measured experimentally using tasks where participants are required to say how much they would be prepared to approach a person whose face they are shown in a photograph. Predominantly, these tasks use rating scales to measure approach desires and have been used with adolescents or adults with the disorder. Ratings that indicate an atypically increased desire to approach unfamiliar people could have important implications if extrapolated to the real world. Within this domain of investigation, there have been a plethora of studies exploring approachability ratings and there are mixed findings regarding the natural of atypicalities. One study reports no atypicality of approach ratings when taking into consideration the emotional understanding of individuals with WS (Porter *et al.*, 2007), a more recent study also supports a relationship between high self-rated willingness to approach and poor affect identification (Järvinen-Pasley *et al.*, 2010). However, other studies indicate abnormally increased approachability toward faces depicting negative expressions (Bellugi *et al.*, 1999; Jones *et al.*, 2000; Martens, Wilson, Dudgeon, & Reutens, 2009) that contrast abnormally high ratings given to faces depicting positive expressions (Frigerio *et al.*, 2006). It is highly likely that emotion perception as well as individual variability of social behavior contributes somewhat to these reported differences. It is also highly likely, however, that these abnormalities of social approach are evident from an early age, with lab-based studies of preschool children with WS showing more desire to interact with a stranger than typically developing children (Dodd, Porter, Peters, & Rapee, 2010).

Atypicality in the way that individuals with WS make socially relevant judgments may also link to abnormalities in the way that individuals with the disorder make social categorizations or form social stereotypes that impact upon their behaviors (see Santos, Meyer-Lindenberg, & Deruelle, 2010). There is little doubt that individuals with WS show atypicalities in the way that they consider familiar and unfamiliar people as approachable and the way that they form social evaluations of other people, however, the exact nature of these atypicalities entails further exploration.

C. ATTENTION TO PEOPLE

Research has indicated that across the lifespan individuals with WS show an extreme interest in looking at people and their faces. This interest is present in infancy (Mervis, Morris, *et al.*, 2003; Mervis, Robinson, *et al.*, 2003) and persists into adolescence and adulthood (Riby & Hancock, 2008).

Research with toddlers who have WS revealed that during encounters with their geneticist nearly all (23 out of 25) showed atypically prolonged gaze toward the geneticist's face (Mervis, Morris, *et al.*, 2003; Mervis, Robinson, *et al.*, 2003). This behavior contrasted that of typically developing infants of the same chronological age. Indeed young infants with the disorder prefer to look at people than to look at objects (Laing *et al.*, 2002). Similarly, adolescents and adults with the disorder tend to fixate on faces in social scenes and movies for significantly longer than typically developing individuals, as evident through tracking eye movements (Riby & Hancock, 2008, 2009). This prolonged facial attention is especially apparent toward the eye region. For example, when attending to faces within a social scene picture adolescents and adults with WS spend over half their time (58% of their face gaze) attending to the eye region and this is significantly longer than typically developing individuals of the same age (36% eye region fixations; Riby & Hancock, 2008). Interestingly, and considering variation across chapters within this text, in the same study children and adolescents with Autism only attended to the eye region for 17% of their face gaze time, indicating differences between neurodevelopmental disorders. Therefore, when engaged in a conversation with a person with WS this high level of fixation upon a person's eyes will manifest itself as extremely intense eye contact that may appear disconcerting and may impact upon the flow of the social interaction. Figure 2 illustrates the attentional fixation patterns of individuals with WS and those with Autism to show cross-syndrome divergence of attention.

D. EMOTIONS

We have described the sensitive, caring, and empathetic nature of individuals with WS (Gosch & Pankau, 1997; Klein-Tasman & Mervis, 2003; Tager-Flusberg & Sullivan, 2000) and it could be hypothesized that these features imply that individuals with the disorder are especially good at deciphering socio-emotive cues from other people. Indeed, some researchers have proposed that individuals with WS are characterized by relative sparing in the domain of theory of mind (the ability to understand feelings and thoughts expressed by other people; Baron-Cohen, 1995). This may be particularly evident when compared to their other intellectual abilities (Baron-Cohen, 1995; Tager-Flusberg, Boshart, & Baron-Cohen, 1998). However, it is widely reported that individuals with WS have difficulties interpreting both simple and complex emotional states from people's faces. We would suggest, in line with the findings of Porter *et al.* (2007) that problems with rating faces appropriately for approachability (as noted previously) are entwined with the ability to correctly infer

Fig. 2. Each small dot represents a fixation being made by a participant as they attended to this frame taken from a movie showing of four adults engaged in conversation. Top image shows the fixation scatters for individuals on the autistic spectrum (of varying degrees of functioning) and bottom image shows the fixation scatters for individuals with WS. For the movie overall, individuals with WS spent significantly longer than typically developing individuals fixating on the face regions and those with autism spent significantly less time than typical attending to the same regions (for full details see Riby & Hancock, 2009).

the emotion shown in a face that is being judged. When individuals with WS have problems making emotion judgments from faces, this may lead to an inappropriate evaluation of approachability. Indeed research has suggested that both children and adults with WS fail to give attentional priority to angry faces (Dodd & Porter, 2010; Santos, Meyer-Lindenberg, & Deruelle, 2010; Santos, Silva, Rosset, & Deruelle, 2010) but that adolescents and adults show increased attentional bias toward happy faces (Dodd & Porter, 2010). These attentional biases may link emotion perception to other aspects of WS (see Santos, Meyer-Lindenberg, & Deruelle, 2010; Santos, Silva, et al., 2010 for a discussion) and indeed emotion perception is likely to play a central role in the WS social phenotype.

In studies of explicit emotion recognition ability, researchers have repeatedly found performance at a level characteristic of individuals with other forms of intellectual difficulty (Gagliardi et al., 2003; Plesa Skwerer, Faja, Schofield, Verbalis, & Tager-Flusberg, 2006; Plesa Skwerer, Verbalis, Schofield, Faja, & Tager-Flusberg, 2006). Problems exist across the developmental spectrum of individuals with the disorder. When discriminating basic expressions of emotion (e.g., happy, sad) from schematic faces children with WS perform at a level comparable to mental age matched typically developing children (Karmiloff-Smith, Klima, Bellugi, Grant, & Baron-Cohen, 1995), even when completing a task that was originally designed for much younger individuals. Slightly more complex tasks, such as sorting basic expressions (Tager-Flusberg & Sullivan, 2000) and recognizing basic expressions from moving faces (Gagliardi et al., 2003), prove even more challenging. It is highly likely that atypicalities in the way that individuals with WS attend to faces (linking to the abnormalities reported above) throughout development play a role in emotion recognition problems (see Porter et al., 2010). Some individuals with WS may spend so long attending to the eye region that they fail to absorb and process information distributed throughout the face which might provide subtle cues of emotion.

For more complex emotions or mental states (e.g., disinterested, worried), individuals with WS can perform relatively competently when choosing from two semantically opposite options for labeling an emotion from the eye region of a face using the "Reading the Mind from the Eyes" task (Baron-Cohen, Wheelwright, & Jolliffe, 1997; Tager-Flusberg et al., 1998). However, when the task is made slightly more difficult with four unrelated answer options (see Baron-Cohen, Wheelwright, & Hill, 2001), adolescents and adults (12–36 years) with WS perform at a level comparable to individuals with other intellectual difficulties (Plesa Skwerer, Verbalis, et al., 2006b). There is certainly no evidence of proficient performance that seems to be suggested from the social qualities associated with the disorder. However, recent research using whole moving faces has indicated that when individuals with WS (8–23 years of age) have the whole face to use for emotion interpretation, as would be the case in everyday social interactions, individuals with WS perform as well as typically developing individuals of comparable age (Riby & Back, 2010). The previous tasks tapping such skills had relied on use of a black and white strip of the face showing the eye region and had not used stimuli that might mirror the types of face cue readily available during a social interaction. Using this type of stimuli may be critical to educating individuals with WS about the interpretation of emotive cues from people's faces and could have a place in the development of potential social skills training programs for this population.

IV. Psychopathology

The social tendencies (as well as the cognitive capabilities) outlined above may mask a complex range of behaviors and personality attributes that accompany a diagnosis of WS. The difficulty forming and maintaining friendships may, in part, be the consequence of the range of social atypicalities reported above (e.g., a tendency to over attend to faces, increased approach toward strangers) but may also be the result of a low tolerance for frustration, excessive chatter, and impulsive behavior (see Dykens & Rosner, 1999). Indeed we have previously noted attention deficits associated with the disorder and parental ratings for children with the disorder indicate severe difficulties with inattention and hyperactivity that mirror those associated with ADHD (Rhodes *et al.*, in press). However, compared to the cognitive profile associated with the disorder, relatively few studies have focused on psychopathology or maladaptive behaviors in WS. In addition, there are even fewer studies that have explored psychopathology in a suitably large cohort.

A. MALADAPTIVE BEHAVIORS

In a large and thorough study of children with WS (4–16 years of age), Leyfer *et al.* (2006) found that just over 64% met the criteria for ADHD. Of this group, the majority met criteria for "ADHD—Inattentive" subtype. Rhodes *et al.* (in press) found that parents of children with WS reported (using the Conners Rating Scale) as severe problems with inattention and hyperactivity as parents of children with ADHD. Therefore, the severity of these symptoms may be substantial and impact upon daily living. Ten of the 11 children were rated as being within the abnormal range for hyperactivity (Rhodes *et al.*, in press). In similar work, Porter *et al.* (2008) found that 33% of children in their sample met criteria for ADHD once their level of general intellectual ability was taken into account. Therefore, there may be great variability between individuals in terms of inattention and hyperactivity but these issues should be considered as a relevant to individuals with the disorder.

Despite a reasonably high likelihood of inattention and hyperactivity in WS, problems of conduct or oppositional behaviors are relatively rarer (Porter *et al.*, 2008). In the same study reported above, parents of only 5 of the 11 children involved in research by Rhodes *et al.* (in press) rated their child as being within the abnormal range for oppositional behavior. Despite this, some interesting patterns seem to be emerging. For example, Porter *et al.* (2008) found that parents of younger children with WS were more likely to report difficulties with externalizing behaviors (such as oppositional and conduct problems) than parents of older children and adults with WS. A pattern of greater externalizing difficulties in younger children than older children and adults is consistent with other neurodevelopmental syndromes, such as ADHD and Asperger syndrome, at least anecdotally (Flom, 2008; Freeman, 2009). Also, females in Porter *et al.*'s cohort tended to be rated as displaying more externalizing behaviors than males, especially in terms of conduct problems. This finding is the opposite pattern to that seen in typical development and (to the best of the authors' knowledge) is not seen in other neurodevelopmental disorders (e.g., see Rucklidge, 2010). It is therefore necessary to explore these issues at an individual level (and take the individual needs into consideration in the design of interventions).

Other types of maladaptive behavior such as behavioral, emotional, and thought regulation difficulties (e.g., impulsivity, low frustration tolerance, obsessive thoughts, preoccupations) are reasonably common and indeed are more common than seen in the typical population. All these problems are likely to be secondary to executive dysfunction (Dodd & Porter, 2009; Gosch & Pankau, 1997; Mobbs *et al.*, 2007; Porter *et al.*, 2008) and therefore interventions targeting executive functioning and attentional control may have a "knock on" effect for these aspects of maladaptive behavior.

B. ANXIETY DISORDERS

The area of psychopathology that has received the most research attention in WS has been the prevalence and phenomenology of anxiety. Although the exact incidence varies, group studies on psychopathological functioning indicate a high incidence of both generalized anxiety disorder (GAD) and specific phobias compared to other genetic syndromes, individuals with intellectual disability, and typically developing controls (Dodd & Porter, 2009; Dykens, 2003; Einfeld *et al.*, 1997; Leyfer *et al.*,

2006). Prevalence has been reported to be as high as 11.8% for GAD and 53.8% for specific phobia (Leyfer *et al.*, 2006).

Linking to other components of behavior that we have already detailed in this chapter, Dodd, Schniering, and Porter (2009) investigated the prevalence of social anxiety in 12–28-year-olds with WS using both parent- and self-report. Despite their striking social behaviors, the incidence of social anxiety was similar to the prevalence observed in typically developing controls and in a clinically anxious comparison group. This suggests that the high level of social approach associated with WS (reported previously) is not due to lowered levels of social anxiety within the population. Dodd *et al.* (2009) also explored the cognitive processes that relate to symptoms of nonsocial anxiety. Findings indicated that thoughts regarding physical threat were higher in WS than in typically developing children. Moreover, the relationship between these thought patterns and symptoms of generalized anxiety paralleled those patterns observed in a clinically anxious comparison group. The finding of a similar relationship between thought patterns and GAD in WS and clinically anxious individuals suggests that mainstream interventions such as cognitive behavior therapy may be beneficial in reducing anxiety symptoms for individuals with WS.

C. DEPRESSIVE DISORDER

Using a DSM-IV diagnostic clinical interview to comprehensively explore a range of psychopathological disorders across the lifespan in WS, Dodd and Porter (2009) found that 14% of their sample ($n=50$) and 25% of their adult group alone met criteria for a depressive disorder. This suggests that, in addition to previously documented difficulties with anxiety and inattention, depressive disorders may also present a significant problem for individuals with WS. This may especially be the case during late adolescence and adulthood, a timeframe when individuals with the disorder may experience challenges of independence versus social isolation. Depressive disorders may pose a significant problem for individuals with WS and their families and should not be overlooked (or diagnostically overshadowed).

V. Interventions

Interventions targeting each of the key areas of cognition and behaviors discussed throughout this chapter will now be considered.

A. INTERVENTIONS TARGETING LANGUAGE

Any intervention targeting language capabilities must be aware of the uneven profile *within* the language domain (that may be unique to WS) and consider how best to utilize the skills that individuals with WS have in relation to areas of more extreme problems. However, almost all individuals with WS would benefit from speech and language therapy, as it would be extremely rare to find an individual with WS with all language abilities at a level expected for their age. Accordingly, the majority of individuals with WS do undergo speech and language therapy (Porter & Coltheart, 2005; Semel & Rosner, 2003). As Mervis and Becerra (2007) point out, language development appears to relate to cognitive development more generally. Given the vast heterogeneity of cognitive ability seen in WS, we argue (along with Mervis & Becerra, 2007) that speech and language therapy should be individually tailored to each individual's profile of relative strengths and weaknesses (indeed this approach would be needed for individuals with any of the disorders of development covered in this text). This process of individuation must consider not only the language profile but an overall neuropsychological profile. Mervis and Becerra (2007) also highlight an important point with regard to potential atypicalities of language development in WS and implications for intervention; unlike in the typical population, referential communicative gestures do not fully develop prior to the rapid onset of expressive vocabulary and, therefore, if language problems are not identified until after the early stages of vocabulary acquisition, speech and language therapists may incorrectly assume that children with WS have successfully mastered referential gesture system. Therefore, interventions for individuals with WS (and indeed individuals with other neurodevelopmental disorders covered in this book) should ignore the milestones expected in typical developmental trajectories and must develop syndrome-specific approaches to the language interventions. Therapists also need to be creative and open to the fact that different, innovative intervention approaches may be required, focusing on the unique profile of proficiencies and deficiencies presented by each child. Indeed if language capabilities are a relative strength for an individual with WS, they should be encouraged to use their language skills in other aspects of cognition; for example, talking through a problem they may be trying to solve (Dykens, Hodapp, & Finucane, 2000). Therefore, individuals may be able to use their language competencies to aid other aspects of cognition and feed into interventions aimed at targeting more problematic skills.

Within an evaluation and intervention approach, standardized language assessments such as the Clinical Evaluation of Language Fundamentals 4

(CELF-4) should be utilized in conjunction with more direct observations. This approach could be applied to any of the disorders covered here. Whilst a standardized evaluation will be useful direct observations that consider behaviors associated with WS will more sensitively identify pragmatic language difficulties and other language problems that relate to executive functioning deficits (e.g., perseverative speech, inappropriate greeting behavior or inappropriate remarks, word finding difficulties, and excessive talkativeness). Other cognitive aspects, such as poor attention and mental fatigue should be considered when structuring any therapy session. Behavior modification (e.g., positive reinforcement) may be particularly beneficial in conjunction with the use of speech therapy to address executive functioning deficits and their impact on language functioning. Therefore, in summary, these considerations can be used to design targeted language interventions for individuals with WS.

B. INTERVENTIONS TARGETING VISUOSPATIAL SKILLS

Many individuals with WS and their families seek treatment from ophthalmologists and/or behavioral optometrists to assist with deficits in visual acuity, visual perception, visual scanning, and visual tracking. Additionally, occupational therapists are often utilized to assist with delayed early motor development, motor planning, and fine-motor difficulties, as well as spatial construction and visuomotor integration deficits. Improvement of handwriting and drawing skills is a common occupational therapy goal for preschool and primary school-aged children with WS. Semel and Rosner (2003) suggest that verbal mediation strategies may also be used to assist with enhancing drawing and handwriting skills, and to assist with other spatial deficits such as topographical disorientation (refer back to the use of verbal mediation for language interventions). This approach may be more beneficial for individuals with WS than those with other disorders (e.g., DS) due to the unique shape of the WS cognitive profile. For example, verbal descriptions can be used to help WS individuals negotiate the position of a pen on a page when forming letters and may also assist with navigating common travel routes.

C. INTERVENTIONS TARGETING ACADEMIC SKILLS

Having considered the academic skills of individuals with WS it is important to think about how interventions may be designed to best accommodate and improve these skills. We will consider issues of literacy and numeracy. First, the typical or atypical development of reading skill has important implications for improving reading and the design of reading interventions for those with WS (Laing, 2002). The methods that are used in mainstream schools and which have been targeted to the specific needs of typically developing children may not be appropriate in many respects if reading develops atypically in WS. Individual variability provides a further challenge to the development of intervention and educational provisions. For example, while some authors argue that individuals with WS are likely to benefit from a whole-word (sight word) approach to learning to read (Menghini, Verucci, & Vicari, 2004), others support a phonic-based approach (Becerra, John, Peregrine, & Mervis, 2008, as cited by Levy, Smith, & Tager-Flusberg, 2003; Mervis, 2009) and it is highly likely that these different approaches would be more or less suitable to some young children with WS than others. Importantly, more in-depth studies are required before we can make firm conclusions regarding reading instruction and reading intervention in WS (for further discussion, see Mervis & John, 2010). Individually tailored teaching and targeted interventions are likely to be the best approach at this stage.

Another important consideration is that neuropsychological abilities (e.g., spatial perception and visual scanning anomalies) may also be implicated in the reading problems associated with WS and can affect both reading and reading comprehension. Different neuropsychological deficits will have a variable impact on these literacy skills. For example, interventions would be very different for a WS child whose spatial abilities were affecting reading, compared to a child whose phonological awareness and working memory deficits have a more significant impact (indeed the tailoring of these interventions may also be very different to that used for individuals with other disorders of development). As a gold standard, literacy instruction and intervention should be considered after a full neuropsychological profile has been established for a particular WS individual. As an example strategy, for children with WS whose spatial deficits are impacting on their reading or spelling, color coding can be used to highlight important differentiating features in visual stimuli such as in letters which are often confused (e.g., "b"/"d", "t"/"f"). Similarly, for children with visual tracking problems, larger font size, using stencils to cover irrelevant text or encouraging finger tracking may assist.

To consider interventions targeting numeracy skills in WS, Ansari *et al.* (2003) found that counting was significantly correlated with verbal abilities but, in contrast, was significantly related to spatial construction abilities in typical development. The authors suggest that individuals with WS rely on their verbal competencies to "bootstrap" or "scaffold" their understanding of numerical principals (in a way that may not be seen in other disorders of development). As such, verbal mediation may be an important tool for numerical interventions in this group, whereby these individuals can take advantage of their verbal skills when attempting to grasp nonverbal and mathematical concepts (see this aspect in the earlier section of this chapter on language interventions). However, one must always keep in mind that verbal comprehension abilities are also often well below chronological age in WS, so individuals are likely to experience difficulty with lengthy verbal mathematical problems. Also, attention, spatial skills, working memory, and executive functioning abilities such as reasoning and problem-solving skills and impulsivity will also impact on their ability to perform numerical calculations and more verbally based numerical problems.

Following the thought that cognitive strengths may be used to intervene with areas of weaknesses, Kwak (2009) designed a mathematical intervention for individuals with WS using music therapy to assist with understanding mathematical concepts. The particular goal was to increase the individuals' motivation to attend to mathematical tasks for longer periods of time. The authors concluded, however, that many aspects of cognition appear to be impacting on mathematical abilities and that different interventions would be required to address each cognitive difficulty. Nevertheless, the authors suggested that their findings did provide evidence supporting the use of music therapy with an intervention program targeting mathematical development for individuals with WS.

Finally, a simple technique that teachers could utilize with their pupils may be to reduce working memory load during mathematical computations (e.g., see Gathercole & Alloway, 2008). Teachers may find that using counters is useful for young children with WS. Additionally, calculators may be used by older individuals. WS individuals should also be encouraged to check over their work and focus on accuracy rather than speed to reduce the impact of impulsivity on their mathematical performance.

It is evident that there are several ways that interventions can be aimed at improving literacy and numeracy skills for individuals with WS. In this section, we highlight that having a full profile of each individual with the disorder is critical for understanding the link between strengths and weaknesses and the possible use of compensatory mechanisms to boost performance decrements. Once teachers have an understanding of the full profile of strengths and weaknesses for an individual they can put in place strategies to increase basic literacy and numeracy skills.

D. INTERVENTIONS TARGETING EXECUTIVE FUNCTIONS AND ATTENTION

Although a small number of studies have made suggestions regarding the possible nature of future interventions targeting attention problems, a scarcity of studies have put these possibilities into practice. These interventions will be covered in subsequent sections of this chapter where we note the relevance of attention and executive functioning to problems of social behavior and maladaptive (e.g., hyperactive) tendencies. In general, environmental modifications such as removing distracting stimuli, sitting the child at the front of the class next to good role models, providing structure and routine, and behavior modification (positive reinforcement) are likely to be most beneficial when it comes to minimizing the impact of attention and executive functioning difficulties on the daily lives of individuals with WS. These approaches to intervention may mirror those used for individuals with any disorder of development that impacts upon this domain of functioning.

E. INTERVENTIONS TARGETING SOCIAL BEHAVIORS

There have been minimal attempts to design interventions targeted at the social behaviors associated with WS. This is likely to be due to the general impression that because individuals with WS tend to "thrive" on social engagement this must be an area of relative strength (certainly in terms of other problems with cognition). However, with clearer understanding of the subtle atypicalities of social functioning that some individuals with the disorder experience (e.g., hyper sociability, atypical approach) and social vulnerability for adolescents and adults with the disorder (Jawaid *et al.*, submitted), there is a timely need to consider how best to target these problems. Such interventions would be particularly beneficial to individuals with the disorder as well as their families.

Interestingly, there are a significant number of interventions aimed at the range of social behavioral problems associated with Autism that may also be beneficial for individuals with WS; with some slight modification to provide a truly individualized and targeted program. Early interventions probing the development of joint attention skills may be useful in probing not only social initiation problems but also the early precursors of language and social engagement (see work with ASD; Kasari, Freeman, & Paparella, 2006). Similarly, work using social stories may provide individuals with WS with the skills they require to make suitable and appropriate evaluations of social approach and to teach social boundaries.

If this intervention includes a consideration of "stranger danger" this would be particularly valuable to the specific type of atypicalities of social approach associated with WS (see work with ASD; e.g., Reichow & Sabornie, 2009; Scattone, Wilczynski, Edwards, & Rabian, 2002).

It may also be useful for intervention programs to take into consideration other aspects of behavior of cognition that may affect, or indeed be affected by, social behaviors. For example, Rhodes *et al.* (2010) propose that executive functioning problems are likely not to be isolated to cognition but are highly likely to extrapolate to social behaviors (e.g., problems inhibiting social responses). In a similar vein, Porter *et al.* (2008) found that WS individuals who showed deficits recognizing basic expressions of emotion were also rated by their parents as showing oppositional and conduct problems. Thus, an intervention to improve emotion recognition in such WS individuals may not only lead to an improved ability to read others emotions and interact socially, but may also lead to improvements in other aspects of behavior.

Finally, it is worth noting that assessments and training programs targeting aspects of approachability would benefit from the use of evaluation methods other than rating scales. Understanding the scoring involved on a rating scale may by particularly difficult for individuals with mild–moderate intellectual difficulties. Therefore, further insights are warranted to identify how best to design suitable interventions or training strategies that focus on issues of "stranger danger" and suitable approaches towards both familiar and unfamiliar people. So taking all this into consideration, much work remains to target the individual needs of children and young people with WS in terms of their social skills problems; with this aim in mind it may subsequently be possible to reduce the prevalence of social isolation and social vulnerability on adults with the disorder.

F. INTERVENTIONS TARGETING PSYCHOPATHOLOGY

We have highlighted that individuals with WS display a range of maladaptive behaviors that impact upon cognition, social behavior, and educational attainment; including distractibility, inattention, inflexibility, impulsivity, low frustration tolerance, atypical activity, and increased fears and anxieties (Semel & Rosner, 2003). As proposed by Porter *et al.* (2008), it is particularly important that the individual profile of skills and deficits is taken into consideration when exploring the range of psychopathological behaviors the individual may exhibit. Although the literature suggests there is no effect of mental age or IQ on psychopathological symptoms or diagnoses in WS (Dodd & Porter, 2009; Dykens & Rosner,

1999, Leyfer *et al.*, 2006; Porter *et al.*, 2008), we propose that the profile of personal cognitive strengths and weaknesses should be considered in explorations and interventions of psychopathology in WS (see also Porter *et al.*, 2008).

Linking to the suggestion of similar attention deficits and behaviors seen in children with WS and ADHD, many families are now turning to pharmaceutical treatments, such as stimulant medication to assist with attention and behavior difficulties, but there is a scarcity of research in this area with regards to WS. Bawden, MacDonald and Shea (1997) conducted a case series of double-blind, placebo-controlled methylphenidate (Ritalin) trials aimed to improve behavior problems and attention in four children with WS. Two of the four children in the methylphenidate group responded favorably, showing decreased impulsivity, less irritability, and improved attention. Similarly, a small case study of two children with WS has shown the possible use of methylphenidate, with a significant improvement of play behavior and with teachers reporting improvements for a range of behaviors (Power, Blum, Jones, & Kaplan, 1997). Following from these two studies and evidence of the overlapping severity of symptoms compared to children with ADHD, Rhodes *et al.* (in press) suggest that it may be possible to implement some of the behavioral, educational, and pharmacological treatments that have previously been useful for children with ADHD, with children who have WS. However, to date we have a lack of evidence to support this implementation and further work is essential in this domain. These interventions should bear in mind that although the observable behaviors of individuals with WS and ADHD may appear to be the similar, it cannot be assumed that the underlying causes are the same. There are many different medications that could be of use to assist with attention and behavior difficulties in WS. It is likely that, as with children who have ADHD, different medications will be of benefit for different children, for some children stimulant medications will have no beneficial effect and, as always, pharmacological interventions should be used in conjunction with psychological intervention to maximize improvements. Importantly though, given the widespread implications executive dysfunction across the lifespan, the design of individually tailored intervention strategies is particularly timely. Indeed these deficits are unlikely to be isolated to cognition and are likely to have more widespread implications for social behaviors (see Porter *et al.*, 2007; Rhodes *et al.*, 2010).

Klein-Tasman and Albano (2007) provide some support for the successful use of traditional methods of psychological intervention in WS. Klein-Tasman and Albano conducted a case study with a young adult who had WS using intensive short-term cognitive–behavioral therapy to treat symptoms associated with obsessive compulsive disorder (OCD) and associated social difficulties. There was an increase in the individual's insight and understanding of his problems and also a decrease in the symptoms associated with the presence of OCD. Also, Phillips and Klein-Tasman (2009) illustrated two case examples of how cognitive behavior intervention could be applied to adolescent females with WS to help them cope more effectively with their anxiety. Therefore, although there has, to date, been a lack of intervention work in this area with individuals who have WS, there may be a range of methods available to clinicians and therapists for use with this population and which may be ideally suited to targeting some of the psychopathologies rated to the disorder.

In summary, there is very little research on the efficacy of psychological, behavioral, and pharmaceutical interventions for treating or modifying behavioral and psychopathological issues, but studies to date show promising albeit beginning support for these methods in reducing the impact of such disorders. Further studies are, however, desperately required in this area.

G. SUMMARY OF INTERVENTIONS

Throughout this chapter, we have aimed to illustrate the diverse ways that WS may impact upon cognition and behavior and the ways that interventions may be purposefully targeted toward the issues relevant to individuals with the disorder. In this section, we summarize the key issues that may enhance the utility of intervention programs for this group. Some of these issues may be generic and relate to intervention programs used with individuals who have other disorders of development (e.g., those discussed in other chapters of this book) whilst some may be more specialized for individuals with WS. Mervis and John (2010) sum up the importance of developing such intervention programs by emphasizing a need "to provide a sound research basis for the development of the educational, social, and behavioral interventions needed for children with WS to have the opportunity to reach their full potential" (page 245).

We would propose that a generic group approach (the "cookie cutter" design) to interventions will not work for many individuals with WS. Throughout this chapter, we have emphasized the critical role of individual variability in shaping our understanding of the phenotypes associated with WS. From language capabilities to visuospatial deficits and abnormalities of behavior there is extreme variation between individuals who have a diagnosis of WS. This individual variability will uniquely shape the profile of cognition for each individual and will subsequently place unique demands upon their intervention requirements. Taking into consideration the full range of abilities and deficits that an individual presents with will be vital to the design and success of any intervention program.

Throughout this chapter, we have also noted the cross-domain implications of behavior, abilities, and deficits associated with the disorder and this issue is also central to the design and implementation of intervention strategies. For example, an individual who has problems with impulsive behaviors may also exhibit more basic deficits of controlling their attention and show a range of executive functioning problems. Therefore, it is critical that we do not consider behaviors in isolation but tackle the whole profile of strengths and weaknesses when considering how best to approach the needs of any individual with WS. These issues promote the use of individually tailored intervention strategies that take a holistic approach to considering the needs to each individual.

Of course, these needs may vary depending on the age; any targeted intervention program should consider the changing demands that are placed upon an individual with WS at different time points across the lifespan. For example, academic interventions may be most suitable early in development whereas social programs to minimize social withdrawal may emerge later when individuals leave school and are no longer interacting regularly with their peers to reduce the possibility of social isolation. Therefore, we should consider the important milestones that an individual with WS may face (e.g., starting school, starting high school, finishing school, starting work). These milestones may be the same or different to individuals who are developing typically but they will allow us to prioritize the needs of individuals with the disorder.

The unique profile of cognitive strengths and weaknesses associated with WS mean that we have the ideal opportunity to utilize the individual capabilities of each person with the disorder in tackling issues that they may find more challenging. For example, if an individual shows strengths at language they may be more likely to benefit from the use of verbal mediation (e.g., talking through a problem as they solve it) more than someone who has less language capability. Similarly, someone with WS who shows an interest and proficiency in music may be able to use music therapy to enhance their cognitive or emotional goals (e.g., increase attention or reduce anxiety). However, once again we need to work at the individual level to assess these issues. An additional demand upon intervention design may also be that strengths and difficulties may change in a single individual over the course of development. Therefore, it is important to use reasonably regular neuropsychological assessments to evaluate both the needs of an individual and their intervention requirements.

Within the WS literature to date, most intervention recommendations for this group involve behavioral observations and the use of clinical judgment; interventions are yet to be fully put to the challenge of empirical scrutiny. There is therefore a real need for further work on possible intervention programs for the range of needs faced by these individuals. This chapter and the scarcity of research evidence that is available for evaluating the efficacy of intervention programs, emphasize a dramatic lack of knowledge in this area.

VI. Conclusions and Future Directions

We have made several key points throughout this chapter that warrant consideration in terms of the current state of knowledge and future research directions. We note that when the cognitive profile of WS is described we can be justified in describing the peaks and valleys of performance and abilities but this must be done alongside the backdrop of a general mild–moderate level of intellectual disability; rather than claims of spared and impaired performance. Additionally, we must consider how each individual varies in their profile of relative strengths and weaknesses, and indeed how development as a whole may vary from a typical pathway. In fact, the profile of relative strengths and weaknesses of cognition should be considered within a developmental framework whereby those strengths/weaknesses may change with age. Finally, when considering the abilities of an individual we should take a holistic approach to consider how their specific cognitive capabilities may relate to wider issues of social functioning or psychopathology; throughout the chapter we have noted the interplay between these characteristics. Throughout the chapter, we have neglected the important role of neural mechanisms underpinning the skills and deficits reported here. It is important that any full profile of individuals with the disorder also considers neural atypicalities that may be implicated in observable behaviors (e.g., frontal lobe dysfunction that is involved in executive functioning abnormalities or amygdala atypicalities that may help shape social functioning).

All of the above issues are relevant to the design of targeted intervention programs that can consider the needs of an individual with WS. To date, the field of WS (in terms of intervention evaluations) is plagued with varied findings from a range of studies using different instruments, different control groups, and small sample sizes. Wider consideration is required to bear in mind the needs of each person with the disorder and how best to use their strengths to combat problem behaviors or cognitive deficits. Empirical evidence using information from varied and

multidisciplinary approaches and taking a longitudinal perspective is critical to advancing knowledge in this area. Ultimately, successfully
implemented intervention strategies could significantly enhance the life
experiences of individuals with WS and their families.

REFERENCES

Annaz, D., Karmiloff-Smith, A., Johnson, M. H., & Thomas, M. S. C. (2009). A cross-syn-
    drome study of the development of holistic face recognition in children with autism,
    Down syndrome, and Williams syndrome. *Journal of Experimental Child Psychology*,
    102, 456–486.
Ansari, D., Donlan, C., Thomas, M., Ewing, S. A., Peen, T., & Karmiloff-Smith, A. (2003).
    What makes counting count? Verbal and visuo-spatial contributions to typical and atypi-
    cal development. *Journal of Experimental Child Psychology*, *85*, 50–62.
Atkinson, J. (2000). *The developing visual brain*. Oxford: Oxford University Press.
Baron-Cohen, S. (1995). *Mindblindness: An essay on autism and theory of mind*. Cambridge:
    MA MIT Press.
Baron-Cohen, S., Wheelwright, S., & Hill, J. (2001). The 'reading the mind in the eyes' test
    revised version: A study with normal adults, and adults with Asperger syndrome or high
    functioning autism. *Journal of Child Psychology and Psychiatry*, *42*, 241–251.
Baron-Cohen, S., Wheelwright, S., & Jolliffe, T. (1997). Is there a 'language of the eyes'?
    Evidence from normal adults and adults with autism or Asperger syndrome. *Visual
    Cognition*, *4*, 311–331 ldren with Williams syndrome using methylphenidate.
Bawden, H. N., MacDonald, W., & Shea, S. (1997). Treatment of children with Williams syn-
    drome with methylphenidate. *Journal of Child Neurology*, *12*, 248–252.
Becerra, A. M., John, A. E., Peregrine, E., & Mervis, C. (2008). *Reading abilities of 9–17-
    year-olds with Williams syndrome: Impact of reading method*. *Poster Presented at the Sym-
    posium on Research in Child Language Disorders*. Madison, WI.
Bellugi, U., Bihrle, A., Jernigan, T. L., Trauner, D., & Doherty, S. (1990). Neuropsychologi-
    cal, neurological, and neuroanatomical profile of Williams syndrome. *American Journal of
    Medical Genetics*, *6*, 115–125.
Bellugi, U., Klima, E. S., & Wang, P. P. (1996). Cognitive and neural development: Clues
    from genetically based syndromes. In D. Magnussen (Ed.), *The lifespan development of
    individuals: A synthesis of biological and psychological perspectives* (pp. 223–243). New
    York: Cambridge University Press. The Nobel Symposium.
Bellugi, U., Lai, Z., & Wang, P. P. (1997). Language, communication, and neural systems in
    Williams syndrome. *Mental Retardation and Developmental Disabilities Research Reviews*,
    3, 334–342.
Bellugi, U., Marks, S., Bihrle, A. M., & Sabo, H. (1988). Dissociation between language and
    cognitive functions in Williams syndrome. In D. Bishop & K. Mogford (Eds.), *Language
    development in exceptional circumstances* (pp. 177–189). Edinburgh: Churchill
    Livingstone.
Bellugi, U., Mills, D., Jernigan, T. L., Hickok, G., & Galaburda, A. (1999). Linking cognition,
    brain structure, and brain function in Williams syndrome. In H. Tager-Flusberg (Ed.),
    Neurodevelopmental disorders (pp. 111–136). Cambridge: MIT Press.

Bellugi, U., Sabo, H., & Vaid, J. (1988). Spatial deficits in children with Williams syndrome. In U. Bellugi (Ed.), *Spatial cognition: Brain bases and development*. Hillsdale, NJ: Erlbaum.

Bellugi, U., Wang, P. P., & Jernigan, T. L. (1994). Williams syndrome: An unusual neuropsychological profile. In S. H. Broman & J. Grafman (Eds.), *Atypical cognitive deficits in developmental disorders: Implications for brain function* (pp. 23–66). Hillsdale: Lawrence Erlbaum Associates.

Benton, A. L., Hamsher, K., Varney, N. R., & Spreen, O. (1983). *Benton test of facial recognition*. New York, NY: Oxford University Press.

Bertrand, J., Mervis, C. B., & Eisenberg, J. D. (1997). Drawing by children with Williams syndrome: A developmental perspective. *Developmental Neuropsychology, 13*, 41–67.

Bihrle, A. M., Bellugi, U., Delis, D., & Marks, S. (1989). Seeing either the forest or the trees: Dissociation in visuospatial processing. *Brain and Cognition, 11*, 37–49.

Bishop, D. V. M., & Baird, G. (2001). Parent and teacher report of pragmatic aspects of communication: Use of the children's communication checklist in a clinical setting. *Developmental Medicine and Child Neurology, 43*, 809–818.

Brock, J. (2007). Language abilities in Williams syndrome: A critical review. *Development and Psychopathology, 19*, 97–127.

Brock, J., Jarrold, C., Farran, E., Laws, G., & Riby, D. M. (2007). Do children with Williams syndrome really have good vocabulary knowledge? Methods for comparing cognitive and linguistic abilities in developmental disorders. *Clinical Linguistics and Phonetics, 21*, 673–688.

Carey, S. (1981). The development of face perception. In G. Davies, H. Ellis & J. Shepherd (Eds.), *Perceiving and remembering faces* (pp. 9–38). London: Academic Press.

Davies, M., Udwin, O., & Howlin, P. (1998). Adults with Williams syndrome. Preliminary study of social, emotional and behavioral difficulties. *British Journal of Psychiatry, 172*, 273–276.

Deruelle, C., Mancini, J., Livet, M., Casse-Perrot, C., & de Schonen, S. (1999). Configural and local processing of faces in children with Williams syndrome. *Brain and Cognition, 41*, 276–298.

Dodd, H. F., Porter, M. A., Peters, G. L., & Rapee, R. M. (2010). Social approach in preschool children with Williams syndrome: The role of the face. *Journal of Intellectual Disability Research, 54*(3), 194–203.

Dodd, H. F., & Porter, M. A. (2009). Psychopathology in Williams syndrome: The effect of Individual differences across the life span. *Journal of Mental Health Research in Intellectual Disabilities, 2*, 89–109.

Dodd, H. F., & Porter, M. A. (2010). I see happy people: attention towards happy but not angry facial expressions in Williams syndrome. *Cognitive Neuropsychiatry, 21*, 1–19.

Dodd, H. F., Schniering, C. A., & Porter, M. A. (2009). Beyond Behavior: Is social anxiety low in Williams syndrome. *Journal of Autism and Developmental Disorders, 39*(12), 1673–1681.

Don, A. J., Schellenberg, E. G., & Rourke, B. P. (1999). Music and language skills of children with Williams syndrome. *Child Neuropsychology, 5*, 154–170.

Donnai, D., & Karmiloff-Smith, A. (2000). Williams syndrome: From genotype through to the cognitive phenotype. *American Journal of Medical Genetics: Seminars in Medical Genetics, 97*(2), 164–171.

Doyle, T. F., Bellugi, U., Korenberg, J. R., & Graham, J. (2004). "Everybody in the world is my friend" hypersociability in young children with Williams syndrome. *American Journal of Medical Genetics, 124*, 263–273.

Dykens, E. M. (2003). Anxiety, fears, and phobias in persons with Williams syndrome. *Developmental Neuropsychology*, *23*, 291–316.

Dykens, E. M., Hodapp, R. M., & Finucane, B. (2000). *Genetics and mental retardation syndromes: A new look at behavior and interventions.* Baltimore: Brookes.

Dykens, E. M., & Rosner, B. (1999). Refining behavioral phenotypes: Personality motivation in Williams and Prada–Willis syndromes. *American Journal of Mental Retardation*, *104*, 158–169.

Einfeld, S. L., Tonge, B. J., & Florio, T. (1997). Behavioral and emotional disturbance in individuals with Williams syndrome. *American Journal on Mental Retardation*, *102*, 45–53.

Elgar, K., & Campbell, R. (2001). Annotation: The cognitive neuroscience of face recognition: Implications for developmental disorders. *Journal of Child Psychology and Psychiatry*, *42*, 705–717.

Ewart, A., Morris, C. A., Atkinson, D., Jin, W., Sternes, K., Spallone, P., et al. (1994). Hemizygosity at the elastin locus in a developmental disorder, Williams syndrome. *Nature Genetics*, *5*, 11–16.

Farran, E. K. (2005). Perceptual grouping ability in Williams syndrome: Evidence for deviant patterns of performance. *Neuropsychologia*, *43*, 815–822.

Farran, E. K., & Jarrold, C. (2003). Visuospatial cognition in Williams syndrome: Reviewing and accounting for the strengths and weaknesses in performance. *Developmental Neuropsychology*, *23*, 173–200.

Farran, E. K., Jarrold, C., & Gathercole, S. E. (2003). Divided attention, selective attention and drawing: Processing preferences in Williams syndrome are dependent on the task administered. *Neuropsychologia*, *41*, 676–687.

Flom, M. N. (2008). Patterns of internalized and externalized behaviors on the BASC-2 PRS in higher functioning children with autism. *Dissertation Abstracts International: Section B: The Sciences and Engineering*, *68*, 4864.

Frank, R. A. (1983). *Speech–language characteristics of Williams syndrome: Cocktail party speech revised. Poster Presented at the Meetings of the American speech–Language Hearing Association.* .

Freeman, M. (2009). Examination of the Asperger syndrome profile in children and adolescents: Behavior, mental health and temperament. *Dissertation Abstracts International: Section B: The Sciences and Engineering*, *70*, 1942.

Frigerio, E., Burt, D. M., Gagliardi, C., Cioffi, G., Martelli, S., Perrett, D. I., et al. (2006). Is everybody always my friend? Perception of approachability in Williams syndrome. *Neuropsychologia*, *44*, 254–259.

Frith, U. (1989). *Autism: Explaining the enigma.* Oxford: Blackwell.

Gagliardi, C., Frigerio, E., Burt, D. M., Cazzaniga, I., Perrett, D. I., & Borgatti, R. (2003). Facial expression recognition in Williams syndrome. *Neuropsychologia*, *41*, 733–738.

Gathercole, S. E., & Alloway, T. P. (2008). *Working memory and learning: A practical guide.* London: Sage Publication.

Gathercole, S. E., & Baddeley, A. D. (1989). Evaluation of the role of phonological STM in the development of vocabulary in children: A longitudinal study. *Journal of Memory and Language*, *28*, 200–213.

Georgopoulos, M. A., Georgopoulos, A. P., Kuz, N., & Landau, B. (2004). Figure copying in Williams syndrome and normal subjects. *Experimental Brain Research*, *157*, 137–146.

Gosch, A., & Pankau, R. (1997). Personality characteristics and behavior problems in individuals of different ages with Williams syndrome. *Developmental Medicine and Child Neurology*, *39*, 527–533.

Haas, B. W., Hoeft, F., Searcy, Y. M., Mills, D., Bellugi, U., & Reiss, A. (2010). Individual differences in social behavior predict amygdala response to fearful facial expressions in Williams syndrome. *Neuropsychologia, 48,* 1283–1288.

Hall, S., DeBernardis, M., & Reiss, A. (2006). Social escape behaviors in children with fragile X syndrome. *Journal of Autism and Developmental Disorders, 36,* 935–947.

Hammond, P., Hutton, T. J., Allanson, J. E., Buxton, B., Campbell, L. E., Clayton-Smith, J., *et al.* (2005). Discriminating power of localized three-dimensional facial morphology. *American Journal of Human Genetics, 77,* 999–1010.

Hoffman, J. E., Landau, B., & Pagani, B. (2003). Spatial breakdown in spatial construction: Evidence from eye fixations in children with Williams syndrome. *Cognitive Psychology, 46,* 260–301.

Howlin, P., Davies, M., & Udwin, O. (1998). Cognitive functioning in adults with Williams syndrome. *Journal of Child Psychology and Psychiatry, 39,* 183–189.

Jarrold, C., Baddeley, A. D., & Hewes, A. K. (1999). Genetically dissociated components of working memory: Evidence from Downs and Williams syndrome. *Neuropsychologia, 37,* 637–651.

Jarrold, C., Hartley, S. J., Phillips, C., & Baddeley, A. D. (2000). Word fluency in Williams syndrome: Evidence for unusual semantic organization. *Cognitive Neuropsychiatry, 5,* 293–319.

Järvinen-Pasley, A., Adolphs, R., Yama, A., Hill, K., Grichanika, M., Reilly, J., *et al.* (2010). Affiliative behavior in Williams syndrome: Social perception and real-life social behavior. *Neuropsychologia, 48,* 2110–2119.

Jawaid, A., Riby, D. M., Owens, J., Kass, J. S., White, S. W., & Schulz, P. E. (submitted). 'Too withdrawn' or 'too friendly': Considering social vulnerability in two neuro-developmental disorders.

Joffe, V. L., & Varlokosta, S. (2007). Language abilities in William syndrome: Exploring comprehension, production and repetition. *Advances in Speech-Language Pathology, 9,* 1–13.

Jones, W., Bellugi, U., Lai, Z., Chiles, M., Reilly, J., Lincoln, A., *et al.* (2000). Hyper-sociability: The social and affective phenotype of Williams syndrome. In M. St. George (Ed.), *Journey from cognition to brain to gene* (pp. 43–71). London: The MIT Press.

Karmiloff-Smith, A. (1997). Crucial differences between developmental cognitive neuroscience and adult neuropsychology. *Developmental Neuropsychology, 13,* 513–524.

Karmiloff-Smith, A. (1998). Development itself is the key to understanding developmental disorders. *Trends in Cognitive Sciences, 2,* 389–398.

Karmiloff-Smith, A., Brown, J. H., Grice, S., & Paterson, S. (2003). Dethroning the myth: Cognitive dissociations and innate modularity in Williams syndrome. *Developmental Neuropsychology, 23,* 227–242.

Karmiloff-Smith, A., Grant, J., Berthound, J., Davies, M., Howlin, P., & Udwin, O. (1997). Language and Williams syndrome: How intact is intact? *Child Development, 68,* 246–262.

Karmiloff-Smith, A., Klima, E., Bellugi, U., Grant, J., & Baron-Cohen, S. (1995). Is there a social module? Language, face processing, and theory of mind in individuals with Williams syndrome. *Journal of Cognitive Neuroscience, 7,* 196–208.

Karmiloff-Smith, A., Scerif, G., & Thomas, M. (2002). Different approaches to relating genotype to phenotype in developmental disorders. *Developmental Psychobiology, 40,* 311–322.

Karmiloff-Smith, A., Thomas, M., Annaz, D., Humphreys, K., Ewing, S., Brace, N., *et al.* (2004). Exploring the Williams syndrome face-processing debate: The importance of building developmental trajectories. *Journal of Child Psychology and Psychiatry, 45*(7), 1258–1274.

Kasari, C., Freeman, S., & Paparella, T. (2006). Joint attention and symbolic play in young children with autism: A randomized controlled intervention study. *Journal of Child Psychology and Psychiatry, 47,* 611–620.

Klein-Tasman, B. P., & Albano, A. M. (2007). Intensive, short-term cognitive-behavioral treatment of OCD-like behavior with a young adult with Williams syndrome. *Clinical Case Studies, 6,* 483–492.

Klein-Tasman, B. P., & Mervis, C. B. (2003). Distinctive personality characteristics of children with Williams syndrome. *Developmental Neuropsychology, 23,* 271–292.

Klein-Tasman, B. P., Mervis, C. B., Lord, C. E., & Phillips, K. D. (2007). Socio-communicative deficits in young children with Williams syndrome: Performance on the autism diagnostic observation schedule. *Child Neuropsychology, 13,* 444–467.

Krajcsi, A., Lukács, A., Igács, J., Racsmány, M., & Pléh, C. (2009). Numerical abilities in Williams syndrome: Dissociating the analogue magnitude system and verbal retrieval. *Journal of Clinical and Experimental Neuropsychology, 31*(4), 439–446.

Kwak, E. (2009). An exploratory study of the use of music therapy in teaching mathematical skills to individuals with Williams syndrome. *Dissertation Abstracts International Section A: Humanities and Social Sciences, 70*(2-A), 509.

Laing, E. (2002). Investigating reading development in atypical populations: The case of Williams syndrome. *Reading and Writing: An Interdisciplinary Journal, 15,* 575–587.

Laing, E., Butterworth, G., Ansari, D., Gsoedl, M., Longhi, E., Panagiotaki, G., *et al.* (2002). Atypical development of language and social communication in toddlers with Williams syndrome. *Developmental Science, 5*(2), 233–246.

Laing, E., Hulme, C., Grant, J., & Karmiloff-Smith, A. (2001). Learning to read in Williams syndrome: Looking beneath the surface of atypical reading development. *Journal of Child Psychology and Psychiatry, 42,* 729–741.

Laws, G., & Bishop, D. V. (2004). Pragmatic language impairment and social deficits in Williams syndrome: A comparison with Down's syndrome and specific language impairment. *International Journal of Language & Communication Disorders, 39,* 45–64.

Leder, H., & Bruce, V. (2000). Inverting line drawings of faces. *Swiss Journal of Psychology, 59,* 159–169.

Lenhoff, H. M., Perales, O., & Hickok, G. (2001). Absolute pitch in Williams syndrome. *Music Perception, 18,* 491–503.

Levitin, D. J., & Bellugi, U. (1998). Music abilities in individuals with Williams syndrome. *Music Perception, 15,* 357–389.

Levitin, D. J., Cole, K., Lincoln, A., & Bellugi, U. (2005). Aversion, awareness, and attraction: Investigating claims of hyperacusis in the Williams syndrome phenotype. *Journal of Child Psychology and Psychiatry, 46,* 514–523.

Levy, Y., Smith, J., & Tager-Flusberg, H. (2003). Word reading and reading related skills in adolescents with Williams syndrome. *Journal of Child Psychology and Psychiatry, 44,* 576–587.

Leyfer, O. T., Woodruff-Borden, J., Klein-Tasman, B. P., Fricke, J. S., & Mervis, C. B. (2006). Prevalence of psychiatric disorders in Williams syndrome. *American Journal of Medical Genetics Part B: Neuropsychiatric Genetics, 141,* 615–622.

Lowery, M. C., Morris, C. A., Ewart, A. K., Brothman, L. J., Zhu, X. L., Leonard, C. O., *et al.* (1995). Strong correlation of elastin deletions, detected by FISH, with Williams syndrome: Evaluation of 235 patients. *American Journal of Human Genetics, 57,* 49–53.

Martens, M. A., Wilson, S. J., Dudgeon, P., & Reutens, D. C. (2009). Approachability and the amygdala: Insights from Williams syndrome. *Neuropsychologia, 47,* 2446–2453.

Martens, M. A., Wilson, S. J., & Reutens, D. C. (2008). Research Review: Williams syndrome: A critical review of the cognitive, behavioral, and neuroanatomical phenotype. *Journal of Child Psychology and Psychiatry, 49*, 576–608.

Mehler, J., Jusczyk, P., Lambertz, G., Halsted, N., Bertoncini, J., & Amiel-Tison, C. (1988). A precursor of language acquisition in young infants. *Cognition, 29*, 143–178.

Menghini, D., Verucci, L., & Vicari, S. (2004). Reading and phonological awareness in Williams syndrome. *Neuropsychologia, 18*, 29–37.

Mervis, C. B. (2009). Language and literacy development of children with Williams syndrome. *Topics in Language Disorders, 29*, 149–169.

Mervis, C. B., & Becerra, A. M. (2007). Language and communication development in Williams syndrome. *Mental Retardation and Developmental Disabilities Research Reviews, 13*, 3–15.

Mervis, C. B., & John, A. E. (2010). Cognitive and behavioral characteristics of children with Williams syndrome: Implications for intervention approaches. *American Journal of Medical Genetics, 154*, 229–248.

Mervis, C. B., & Klein-Tasman, B. P. (2000). Williams syndrome: Cognition, personality, and adaptive behavior. *Mental Retardation and Developmental Disabilities Research Reviews, 6*, 148–158.

Mervis, C. B., Morris, C. A., Bertrand, J., & Robinson, B. F. (1999). Williams syndrome: Findings from an integrated program of research. In H. Tager-Flusberg (Ed.), *Neurodevelopmental disorders: contributions to a new framework from the cognitive neurosciences* (pp. 65–110). Cambridge, MA: MIT Press.

Mervis, C. B., Morris, C. A., Klein-Tasman, B. P., Bertrand, J., Kwitny, S., Appelbaum, L. G., et al. (2003a). Attentional characteristics of infants and toddlers with Williams syndrome during triadic interactions. *Developmental Neuropsychology, 23*, 243–268.

Mervis, C. B., & Robinson, B. F. (2000). Expressive vocabulary ability of toddlers with Williams syndrome or Down syndrome: A comparison. *Developmental Neuropsychology, 17*, 111–126.

Mervis, C. B., Robinson, B. F., Bertrand, J., & Morris, C. A. (2000). The Williams syndrome cognitive profile. *Brain and Cognition, 44*, 604–628.

Mervis, C. B., Robinson, B. F., Rowe, M. L., Becerra, A. M., & Klein-Tasman, B. P. (2003b). Language abilities of individuals with Williams syndrome. *International Review of Research in Mental Retardation, 27*, 35–81.

Mobbs, D., Eckert, M. A., Mills, D., Korenberg, J., Bellugi, U., Galaburda, A. M., et al. (2007). Frontostriatal dysfunction during response inhibition in Williams syndrome. *Biological Psychiatry, 62*, 256–261.

Mondloch, C. J., Le Grand, R., & Maurer, D. (2002). Configural face processing develops more slowly than featural face processing. *Perception, 31*, 553–566.

Morris, C. A., Demsey, S. A., Leonard, C. O., Dilts, C., & Blackburn, B. L. (1988). Natural history of Williams syndrome: Physical characteristics. *Journal of Pediatrics, 113*, 318–326.

Morris, C. A., & Mervis, C. B. (1999). Williams syndrome. In S. Goldstein & C. R. Reynolds (Eds.), *Handbook of neurodevelopmental and genetic disorders in children* (pp. 555–590). New York: Guilford.

O'Hearn, K., & Landau, B. (2007). Mathematical skill in individuals with Williams syndrome: Evidence from a standardized mathematics battery. *Brain and Cognition, 64*, 238–246.

O'Hearn, K., & Luna, B. (2009). Mathematical skills in Williams syndrome: Insight into the importance of underlying representations. *Developmental Disabilities Research Reviews, 15*(1), 11–20.

Pani, J. R., Mervis, C. B., & Robinson, B. F. (1999). Global spatial organization by individuals with Williams syndrome. *Psychological Science*, *10*(5), 453–461.

Paterson, S. J., Girelli, L., Butterworth, B., & Karmiloff-Smith, A. (2006). Are numerical impairments syndrome specific? Evidence from Williams syndrome and Down's syndrome. *Journal of child Psychology and Psychiatry*, *47*, 1990 204.

Phillips, K., & Klein-Tasman, B. (2009). Mental health concerns in Williams syndrome: Intervention considerations and illustrations from case examples. *Journal of Mental Health Research in Intellectual Disabilities*, *2*, 110–133.

Philofsky, A., Fidler, D. J., & Hepburn, S. (2007). Pragmatic language profiles of school-age children with autism spectrum disorders and Williams syndrome. *American Journal of Speech Language Pathology*, *16*, 368–380.

Pinker, S. (1999). *Words and rules*. London: Weidenfeld & Nicolson.

Plesa Skwerer, D., Faja, S., Schofield, C., Verbalis, A., & Tager-Flusberg, H. (2006a). Perceiving facial and vocal expressions of emotion in individuals with Williams syndrome. *American Journal on Mental Retardation*, *111*, 15–26.

Plesa Skwerer, D., Verbalis, A., Schofield, C., Faja, S., & Tager-Flusberg, H. (2006b). Social–perceptual abilities in adolescents and adults with Williams syndrome. *Cognitive Neuropsychology*, *23*, 338–349.

Porter, M. A., & Coltheart, M. (2005). Cognitive heterogeneity in Williams syndrome. *Developmental Neuropsychology*, *27*, 275–306.

Porter, M. A., & Coltheart, M. (2006). Global and local processing in Williams, autistic and Down syndrome: Perception, attention and construction. *Developmental Neuropsychology*, *30*(3), 771–789.

Porter, M. A., Coltheart, M., & Langdon, R. (2007). The neuropsychological basis of hypersociability in Williams and Down syndrome. *Neuropsychologia*, *45*, 2839–2849.

Porter, M.A., & Dodd, H. (in press). A Longitudinal study of cognitive abilities in Williams syndrome. *Developmental Neuropsychology*.

Porter, M. A., Dodd, H., & Cairns, D. (2008). Psychopathological and behavior impairments in Williams-Beuren syndrome: The influence of gender, chronological age and cognition. *Child Neuropsychology*, *15*(4), 359–374.

Porter, M. A., Shaw, T. A., & Marsh, P. J. (2010). An unusual attraction to the eyes in Williams–Beuren syndrome: A manipulation of facial affect while measuring face scanpaths. *Cognitive Neuropsychiatry*, *29*, 1–26.

Power, T. J., Blum, N. J., Jones, S. M., & Kaplan, P. E. (1997). Brief report: Response to methylphenidate in two children with Williams syndrome. *Journal of Autism and Developmental Disorders*, *27*, 79–87.

Reichow, B., & Sabornie, E. J. (2009). Brief report: Increasing verbal greeting initiations for a student with autism via a social story intervention. *Journal of Autism and Developmental Disorders*, *39*, 1740 43.

Reilly, J., Harrison, D., & Klima, E. S. (1995). Emotional talk and talk about emotions. *Genetic Counseling*, *6*, 158–159.

Reilly, J., Klima, S., & Bellugi, U. (1990). Once more with feeling: Affect and language in atypical populations. *Development and Psychopathology*, *2*, 367–391.

Reilly, J., Losh, M., Bellugi, U., & Wulfeck, B. (2004). "Frog, where are you?" Narratives in children with specific language impairment, early focal brain injury, and Williams syndrome. *Brain and Language*, *88*, 229–247.

Rhodes, S. M., Riby, D. M., Park, J., Fraser, E., & Campbell, L. E. (2010). Neuropsychological functioning and executive control in Williams syndrome. *Neuropsychologia*, *48*, 1216–1226.

Rhodes, S., Riby, D. M., Matthews, K., & Coghill, D. R. (in press). ADHD and Williams syndrome: Shared behavioral and neuropsychological profiles. *Journal of Clinical and Experimental Neuropsychology*.

Riby, D. M., & Back, E. (2010). Can individuals with Williams syndrome interpret mental states from moving faces? *Neuropsychologia*, *48*, 1914–1922.

Riby, D. M., Doherty-Sneddon, G., & Bruce, V. (2009). The eyes or the mouth? Feature salience and unfamiliar face processing in Williams syndrome and autism. *Quarterly Journal of Experimental Psychology*, *62*, 189–203.

Riby, D. M., & Hancock, P. J. B. (2008). Viewing it differently: Social scene perception in Williams syndrome and autism. *Neuropsychologia*, *46*, 2855–2860.

Riby, D. M., & Hancock, P. J. B. (2009). Looking at movies and cartoons: Eye-tracking evidence from Williams syndrome and autism. *Journal of Intellectual Disability Research*, *53*, 169–181.

Rose, F. E., Lincoln, A. J., Lai, Z., Ene, M., Searcy, Y. M., & Bellugi, U. (2006). Orientation and affective expression effects on face recognition in Williams syndrome and autism. *Journal of Autism and Developmental Disorders*, *37*, 513–522.

Rosner, B. A., Hodapp, R. M., Fidler, D. J., Sagun, J. N., & Dykens, E. M. (2004). Social competence in persons with Prader–Willi, Williams and Down's syndrome. *Journal of Applied Research in Intellectual Disability*, *17*, 209 17.

Rossen, M. L., Klima, E. S., Bellugi, U., Bihrle, A., & Jones, W. (1996). Spared face processing in Williams syndrome. In R. Tannock (Ed.), *Language, learning and behavior disorders: Developmental, biological, and clinical perspectives*. New York: Cambridge University Press.

Rowe, M. L., Peregrine, E., & Mervis, C. B. (2005). *Communicative development in toddlers with Williams syndrome*. Atlanta, GA: Society for Research in Child Development.

Rucklidge, J. J. (2010). Gender differences in attention-deficit/hyperactivity disorder. *Psychiatric Clinics of North America*, *33*, 357–373.

Santos, A., Meyer-Lindenberg, A., & Deruelle, C. (2010a). Absence of racial, but not gender, stereotyping in Williams syndrome children. *Current Biology*, *20*(7), 307–308.

Santos, A., Silva, C., Rosset, D., & Deruelle, C. (2010b). Just another face in the crowd: Evidence for decreased detection of angry faces in children with Williams syndrome. *Neuropsychologia*, *48*, 1071–1078.

Scattone, D., Wilczynski, S. M., Edwards, R. P., & Rabian, B. (2002). Decreasing disruptive behaviors of children with autism using social stories. *Journal of Autism and Developmental Disorders*, *32*, 535 43.

Scerif, G., Cornish, K., Wilding, J., Driver, J., & Karmiloff-Smith, A. (2004). Visual search in typically developing toddlers and toddlers with fragile X or Williams syndrome. *Developmental Science*, *7*, 116–130.

Searcy, Y. M., Lincoln, A., Rose, F., Klima, E., Bavar, N., & Korenberg, J. R. (2004). The relationship between age and IQ in adults with Williams syndrome. *American Journal on Mental Retardation*, *109*, 231–236.

Semel, E., & Rosner, S. R. (2003). *Understanding Williams syndrome: Behavioral Patterns and Interventions* (pp. 16–63). New Jersey: Lawrence Erlbaum Associates, Inc.

Singer-Harris, N. G., Bellugi, U., Bates, E., Jones, A., & Rossen, M. L. (1997). Contrasting profiles of language development in children with Williams and Down syndromes. In D. J. Thal & J. S. Reilly (Eds.), *Origins of language disorders Developmental Neuropsychology 13*, (pp. 345–370). (Special Issue).

Stojanovik, V. (2006). Social interaction deficits and conversational inadequacy in Williams syndrome. *Journal of Neurolinguistics*, *19*, 157–173.

Stojanovik, V., Perkins, M., & Howard, S. (2004). Williams syndrome and specific language impairment do not support claims for developmental double dissociations and innate modularity. *Journal of Neurolinguistics, 17*, 403–424.

Stojanovik, V., & van Ewijk, L. (2008). Do children with Williams syndrome have unusual vocabularies? *Journal of Neurolinguistics, 21*, 18–34.

Tager-Flusberg, H., Boshart, J., & Baron-Cohen, S. (1998). Reading the windows to the soul: Evidence of domain-specific sparing in Williams syndrome. *Journal of Cognitive Neuroscience, 10*, 631–639.

Tager-Flusberg, H., Plesa Skwerer, D., Faja, S., & Joseph, R. M. (2003). People with Williams syndrome process faces holistically. *Cognition, 89*, 11–24.

Tager-Flusberg, H., Plesa-Skwerer, D., & Joseph, R. (2006). Model syndromes for investigating social cognitive and affective neuroscience: A comparison of autism and Williams syndrome. *Social Cognitive and Affective Neuroscience, 1*, 175–182.

Tager-Flusberg, H., & Sullivan, K. (2000). A componential view of theory of mind: Evidence from Williams syndrome. *Cognition, 76*, 59–89.

Tanaka, J. W., & Farah, M. J. (1993). Parts and wholes in face recognition. *Quarterly Journal of Experimental Psychology, 46*, 225–245.

Temple, C. M. (2003). Deep dyslexia in Williams syndrome. *Journal of Neurolinguistics, 16*, 457–488.

Temple, C. M., Almazan, M., & Sherwood, S. (2002). Lexical skills in Williams syndrome: A cognitive neuropsychological analysis. *Journal of Neurolinguistics, 15*, 463–495.

Temple, C., & Clahsen, H. (2002). How connectionist simulations fail to account for developmental disorders in children. *Behavioral and Brain Sciences, 25*, 769–770.

Thomas, M. S. C., Karaminis, T. N., & Knowland, V. C. P. (2010). What is typical language development? *Language Learning and Development, 6*, 162–169.

Tomc, S. A., Williamson, N. K., & Pauli, R. M. (1990). Temperament in Williams syndrome. *American Journal on Medical Genetics, 36*, 345–352.

Udwin, O. (1990). A survey of adults with Williams syndrome and idiopathic infantile hypercalcaemia. *Developmental Medicine and Child Neurology, 32*, 129–141.

Udwin, O., Davies, M., & Howlin, P. (1996). A longitudinal study of cognitive abilities and educational attainment in Williams syndrome. *Developmental Medicine and Child Neurology, 38*, 1020–1029.

Udwin, O., Yule, W., & Martin, N. (1987). Cognitive abilities and behavioral characteristics of children with ideopathic infantile hypercalcaemia. *Journal of Child Psychology and Psychiatry, 28*, 297–309.

van der Geest, J., Lagers-van Haselen, G., & Frens, M. (2006). Saccade adaptation in Williams–Beuren syndrome. *Investigative Ophthalmology and Visual Science, 47*, 1464–1468.

Van Herwegen, J., Ansari, D., Xu, F., & Karmiloff-Smith, A. (2008). Small and large number processing in infants and toddlers with Williams syndrome. *Developmental Science, 11*, 637–643.

Vicari, S., Bellucci, S., & Carlesimo, G. A. (2005). Visual and spatial long-term memory: Differential patterns of impairment in Williams and Down syndromes. *Developmental Medicine and Child Neurology, 47*, 305–311.

Vicari, S., Bellucci, S., & Carlesimo, G. A. (2006a). Evidence from two genetic syndromes for the independence of spatial and visual working memory. *Developmental Medicine and Child Neurology, 48*, 126–131.

Vicari, S., Bellucci, S., & Carlesimo, G. A. (2006b). Short-term memory deficits are not uniform in Down syndrome and Willaims syndromes. *Neuropsychology Review, 16*, 87–94.

Wang, P. P., & Bellugi, U. (1994). Evidence from two genetic syndromes for a dissociation between verbal and visual–spatial short-term memory. *Journal of Clinical and Experimental Neuropsychology, 16,* 317–322.

Wang, P. P., Doherty, S., Rourke, S. B., & Bellugi, U. (1995). Unique profile of visuo-perceptual skills in a genetic syndrome. *Brain and Cognition, 29,* 54–65.

Warrington, E. K. (1984). *Warrington recognition memory test.* Windsor, England: Nfer-Nelson Publishing.

Williams, J., Barrat-Boyes, B., & Lowe, J. (1961). Supravalvar aortic stenosis. *Circulation, 24,* 1311–1381.

Yin, R. K. (1969). Looking at upside down faces. *Journal of Experimental Psychology, 81,* 141–145.

Young, A. W., Hellawell, D., & Hay, D. C. (1987). Configural information in face perception. *Perception, 16,* 747–759.

Zitzer-Comfort, C., Doyle, T., Masataka, N., Korenberg, J., & Bellugi, U. (2007). Nature and nurture: Williams syndrome across cultures. *Developmental Science, 10*(6), 755–762.

FRAGILE X SYNDROME AND ASSOCIATED DISORDERS

Kim M. Cornish, Kylie M. Gray, and Nicole J. Rinehart

CENTRE FOR DEVELOPMENTAL PSYCHIATRY AND PSYCHOLOGY, SCHOOL OF
PSYCHOLOGY AND PSYCHIATRY, FACULTY OF MEDICINE, NURSING AND HEALTH
SCIENCES, MONASH UNIVERSITY, MELBOURNE, AUSTRALIA

I. Introduction

The past decades have witnessed staggering advances in the fields of molecular genetics, cognitive neuroscience, neuropsychiatry, and brain imaging. Collectively, these findings have pushed forward a new generation of research aimed at exploring the dynamic interplay between gene expression, developmental brain pathways, and neurocognitive *profiles* beginning in infancy and moving across the lifespan. These new discoveries have been facilitated by advances on several fronts. New methods are now available for viewing brain activity in real time, there is

Advances in Child Development and Behavior
Patricia Bauer : Editor

expanding information on the complexities of individual genes and gene-
environment interactions, analysis of the domain processes included under
the broad umbrella of "cognition" has become more finer-tuned, and inge-
nious methods have been advanced for measuring typical and atypical
development of these processes. In essence, this research provides a plat-
form to elucidate the complex journey from cell to systems thus allowing a
more precise clinical diagnosis alongside a much clearer understanding of
the behavioral phenotype as it develops across childhood and into adulthood. Most important, these new findings push forward targeted clinical
and educational interventions that recognize disorder-specific strengths
and challenges at major stages of developmental transitions, for example,
from preschool to primary school and from primary school to high school
and then into the workplace.

Fragile X syndrome (FXS) is a well documented neurodevelopmental
disorder but it is rarely defined by its clinical features alone. Although still
lacking consensus, recent estimates indicate a frequency of approximately
1 per 2500 children world-wide will be affected by FXS (Hagerman, 2008).
In some (but certainly not all) children, there is a characteristic constellation of physical features that include an elongated face, large prominent
ears, and forehead, and in males, postpubertal macroorchidism (Cornish,
Levitas, & Sudhalter, 2007; Lachiewicz, Dawson, & Spiridigliozzi, 2000).
More subtle features can include narrow intereye distance, a highly arched
palate of the mouth, and hyperextensible joints. However, the wide
variability in manifestation in both boys and girls makes a diagnosis based
on physical features alone almost impossible. It is precisely because of their
relatively "normal" appearance that many affected children are not diagnosed with FXS until relatively late in their development. Undoubtedly,
the most defining feature, especially in boys, is developmental delay and
the resulting cognitive–behavioral phenotype, most notably the attentional
control difficulties, language impairments, and autistic-like features that
can accompany the syndrome from very early in development (Cornish,
Scerif, & Karmiloff-Smith, 2007; Cornish, Turk, & Hagerman, 2008).
Of significant interest is that it is caused by a single gene being switched
off on the X chromosome resulting in a characteristic *profile* that includes
developmental delay alongside chronic and pervasive attention and executive function difficulties. Furthermore, this same gene is one of the few
known genetic causes of autism, one of the most prevalent and debilitating
of childhood psychiatric disorders.

II. Genetic Profile

An emerging family of DNA mutations known as trinucleotide repeat expansions is responsible for causing a range of cognitive and clinical consequences such as Huntington disease, Friedreich's ataxia, and FXS. The fragile X mental retardation 1 (*FMR1*) gene is a fascinating gene located at the long arm of the X chromosome. This gene contains a cytosine, guanine, guanine (CGG) triplet repeat region that when expanded can result in a continuum of fragile X disorders. Normal CGG repeat sizes correspond to between 7 and 55 repeats, with 30 repeats being the most common. When expanded to >200 repeats (large expansion) the *FMR1* gene is turned off leading to the lack of the fragile X mental retardation protein (*FMRP*), and results in the neurodevelopmental disorder known as FXS. Because FMRP is involved in normal brain development, through its impact on synaptic formation and function, the absence of FMRP results in the characteristic intellectual impairment and cognitive profile associated with this disorder and represents one of the few known single gene causes of autism.

Due to genetic variation in the form of *X-inactivation* (when one of the two X chromosomes remains inactive and the other active) girls with FXS, compared to boys, produce a broader range of cognitive abilities and have IQ's ranging from moderate to the normal range. In contrast, this is not an issue of concern in FXS males whose impairment, without the protection of X-inactivation, shows greater severity. For this reason, we will focus on boys and girls separately.

Most recently, interest has focused on more common, medium size expansions between 55 and 200 CGG repeats (referred to "carrier status"). This research is especially important given the relative frequency of these expansions in the general population calculated as 1 in 130 to 250 females and 1 in 260 to 800 males (Song, Lee, Li, Koo, & Jung, 2003). Until recently carriers were believed to be "phenotypic free", that is without any known cognitive deficits. However, there is a now well documented subtle profile of cognitive strengths and weaknesses, notably in males, that can mirror those found in those individuals with the large CGG expansion (FXS). See Cornish, Turk, *et al.* (2008) for a review of these findings.

III. Neural Profile

Alongside a greater understanding of the genetic underpinnings of FXS, an increasing number of studies have begun to explore how the absence of FMRP impacts on *early brain development*. This research has been facilitated by a variety of new methods that have increasingly allowed neuroscientists to obtain high-resolution images of the structure and activity of the brain. These brain images can capture "snapshots" of brain activity in order to determine which areas are active during task performance. These innovative findings that have begun to enhance our understanding of the typical and atypical developing brain and the complex networks that drive brain maturation across the lifespan.

In typically developing children, the findings from a wealth of new data, made possible by newer imaging technologies, demonstrate the feasibility of imaging the typical brain. The work of Casey and colleagues is testament to the high caliber of research currently being undertaken in this field and the possibilities of this research to push forward our understanding the role of genes on brain development (e.g., Casey, Soliman, Bath, & Glatt, 2010). However, there is still some way to go in providing comparable data of atypical brain development in children with significant cognitive impairment. The most successful studies have tended to use participants with relative higher IQs within the borderline to normal range (therefore skewing the population sample) and there is also a tendency for studies to incorporate older childhood and adolescent samples rather than younger age groups. There are obvious reasons for this but by focusing solely on later outcomes we may be tapping only the end-state rather than exploring age changes in brain maturation over the course of development. In FXS, structural and functional neuroimaging studies highlight a vulnerability of specific brain regions in males and females. For example, there is a decreased size of the posterior vermis of the cerebellum (Mostofsky *et al.*, 1998; Reiss, Aylward, Freund, Joshi, & Bryan, 1991). Other brain areas whose function is affected by FMR1 status include the caudate nucleus (Eliez, Blasey, Freund, Hastie, & Reiss, 2001) and the hippocampus (Kates, Abrams, Kaufmann, Breiter, & Reiss, 1997; Reiss, Lee, & Freund, 1994). Findings of several studies established a correlation between identified structural abnormalities and the degree of cognitive impairment. For example, posterior vermis volumes were found to be positively correlated with performance on specific measures of intelligence, visual–spatial ability, and executive function suggesting a role of this structure in determining performance on these tasks (Mostofsky *et al.*, 1998).

Alongside structural imaging studies, functional magnetic resonance imaging (fMRI) has been the neuroimaging technique of choice for investigating brain structure/function associations in both typically and atypical development. However, studies employing fMRI to investigate FXS are to date limited to female participants (e.g., Kwon *et al.*, 2001). Nevertheless, findings have already demonstrated potential in defining the underlying neural etiology of atypical cognitive functioning associated with the disorder, and in some instances as a function of FMRP expression. For example, both Rivera, Menon, White, Glaser and Reiss (2002) and Kwon *et al.* (2001) demonstrated that reduced parietal activation recorded in adults with FXS was related to *FMR1* protein expression, with possible implications for the visual–spatial and attentional control difficulties reported at the cognitive level. In addition, Tamm, Menon, Johnston, Hessl and Reiss (2002) used fMRI to demonstrate that deficits in cognitive interference during a counting Stroop task may be the result of atypical recruitment of fronto-parietal brain regions. However, the majority of published research has tended to focus on females with a skew toward later adolescence/early adulthood. The reason for this is that FXS females are much less intellectually impaired than their male counterparts and are therefore more able to perform a broader range of cognitive tasks using this technique.

To date, Reiss and colleagues provide the only examples of fMRI studies conducted in both young males and females with FXS with a specific focus on attention control. Using a traditional Go–No Go paradigm, requiring participants to view a series of letters and respond with a key press to every letter except the letter X for which they had to withhold a response, findings indicate that fronto-striatal regions, known to be involved in response inhibition, are especially affected in FXS irrespective of gender. For example, Hoeft *et al.* (2007) compared performance in FXS male adolescents (mean age 15.4 years) and two control groups, an IQ matched developmental delayed group and a typically developing group matched on chronological age. Their findings are noteworthy in two respects: first, Go–No Go performance by FXS males, unlike that of controls males, was not associated with increased activation in the *right* fronto-striatal regions. Second, successful performance was instead associated with increased activation levels in the *left* fronto-striatal network. The pattern of these findings led the authors to make the tantalizing conclusion that response inhibition in FXS may be guided by compensatory processes brought about by a complex interaction between the effects of the *FMR1* gene on early brain maturation, with particular vulnerability in the fronto-striatal network. A similar prefrontal dysfunction has also observed in a study of females with FXS (mean age 15.9 years) using the same Go–No Go paradigm (Menon, Leroux, White, & Reiss, 2004). Undoubtedly, these intriguing findings await further exploration but they clearly demonstrate that at least by adolescence, FXS is associated with anomalous brain development in regions that involve core attention components, in this case attentional control. Future studies will undoubtedly explore brain regions in other cognitive domains.

IV. Cognitive Profile

A. ISSUES OF COMPLEXITY

A core issue of complexity in research that includes defining disorder-specific profiles in children with intellectual impairment is to what extent are (a) behaviors more dependent on the overall degree of impairment, that is, disorder-general deficits no matter what the specific cause (e.g., processing speed differences or low IQ) and (b) which behaviors reflect impairment unique to a particular disorder (syndrome-specific) and/or cognitive domain, that is, domain-specific deficits (such as inhibitory control difficulties or difficulties in face processing) (see Cornish, Bertone, Kogan, Scerif, & Chaudhuri, 2010 for a more detailed discussion). Cornish in a 10-year collaboration with Wilding, Scerif, and Karmiloff-Smith, have exquisitely teased apart the cognitive phenotypes across three genetic disorders of mental retardation: FXS, Williams syndrome (a disorder that results from a microdeletion on one copy of chromosome 7 involving between 25 and 30 genes), and Down syndrome (DS) (a disorder that results from a trisomy of chromosome 21). Together this research has been able to further clarify which behaviors across multiple disorders are more dependent on the overall degree of intellectual impairment, no matter what the specific cause, and which reflect impairment *unique* to a particular genetic disorder. It is therefore critical that research incorporates *cross-syndrome perspectives* to allow a more finer-tuned disorder-specific profile to emerge. In the section below, we will demonstrate how illuminating this approach has been in elucidating the unique nature of cognitive deficits in FXS.

A second issue of complexity relates to the use of *standardized* versus *experimentally* driven paradigms to tap cognitive functioning in atypical populations. All too frequently researchers in this field rely on measures that assess global cognitive functions usually through IQ batteries initially developed for determining the range of functioning in the normal population. As a result, they can mask important yet subtle profiles. For example, standardized tasks are specifically designed to be appropriate in their demands and level of difficulty for a typically developing population. Participants are assumed to be able to understand the instructions, to be motivated to carry out the tasks, to focus for the full duration of the task, and to match various other assumptions. There is no certainty that such assumptions hold when the task is given to participants with a neurodevelopmental disorder; poor performance can occur for a variety of reasons in such cases, not only the inadequacy of the component process or processes that are engaged by the task, and the task may simply be too difficult for the target population. Conversely, a task that is appropriate for a group with cognitive delay may be too easy for a typically developing group. In contrast, experimental measures are developed to be appropriate for developmentally disordered groups such that floor effects can be avoided in those with cognitive delay and also sufficiently sensitive to avoid ceiling effects in the typically developing control groups. There is an emerging literature providing excellent published examples of different types of novel experimental paradigms that include *number* (Ansari, Donlan, & Karmiloff-Smith, 2007), *attention* (e.g., Cornish, Munir, & Wilding, 2001a; Scerif, Cornish, Wilding, Driver, & Karmiloff-Smith, 2004), *working memory* (e.g., Purser & Jarrold, 2005), and *face processing* (e.g., Karmiloff-Smith *et al.*, 2004). We argue that the development of novel experimental paradigms that can tease apart subtle features of performance within and between cognitive domains will facilitate a greater understanding of disorder-specific strengths and difficulties that more generalized measure of cognitive function struggle to isolate.

A third issue is the critical role of *development* itself in producing phenotypic profiles. The pioneering work of Karmiloff-Smith (e.g., Karmiloff-Smith, 2009; Karmiloff-Smith *et al.*, 1998) set the stage for a new generation of research that has focused on charting the developmental trajectories across different neurodevelopment disorders beginning in late infancy and moving across childhood into early adulthood. One of the most prevailing assumptions is that cognitive development in atypical populations represents a static, somewhat "frozen" trajectory of proficiencies and deficiencies that rarely change with increasing age. Thus, one would expect to observe no or minimal age changes in performance across development. However, this assumption has been challenged by recent findings from a number of developmental studies that have shed further light on the possible dynamic role of development in shaping disorder-specific profiles from childhood all the way through to adulthood (see Karmiloff-Smith, 2009, for theoretical perspectives;Thomas *et al.*, 2009). In the section below, we highlight some exciting studies that have incorporated a cross-syndrome perspective in order to identify disorder-specific profiles alongside some emerging findings that have used a developmental approach to address the question of whether cognitive performance in FXS is associated with developmental freeze or developmental change.

B. PROFILE

In FXS, X-linkage means that males are especially vulnerable to the full effects of the condition at the brain and cognitive levels. In boys, almost all present with IQ's within the mild–severe range of impairment with profiles emerging as young as 3 years of age (Skinner *et al.*, 2005). In girls, there is a much broader profile with some girls showing only subclinical learning disabilities (Bennetto & Pennington, 2002) whilst approximately 25% display more significant cognitive impairment (most with mild intellectual disability and rare individuals with moderate intellectual disability) similar in profile to boys with FXS. The X-inactivation status of the FXS female is seen as the major contributor to the heterogeneity of intellectual disability and the broad range of cognitive deficits. However, a decade of research has confirmed that FXS is not defined by the degree of intellectual impairment but rather by a unique "profile" of cognitive strengths and difficulties that differentiate FXS from other neurodevelopmental disorders (e.g., autism, DS, Williams syndrome). A key discovery has been to demonstrate the importance of looking beyond the general effects of developmental delay on intellectual functioning in order to identify the distinct pathways and processes that represent the FXS cognitive profile.

Early studies in the 1980s reported findings that began to explore cognitive functioning using traditional IQ tests to examine potential discrepancies between verbal and nonverbal performance in FXS. These early findings set the scene for research programs beginning in the 1990s and continuing to the present day, with which researchers began to unravel more finely-tuned profiles of cognitive dysfunction—more "skill-specific" rather than "global" in nature (see Cornish, Sudhalter, & Turk, 2004 for a review of these changes in perspectives.) More specifically, by using novel experimental measures that are designed for children with differing levels of intellectual ability and with a focus on delineating performance across a single cognitive domain, such as visuo-spatial or attention, findings revealed unique profiles that differentiate neurodevelopmental disorders from each other and from typically developing children.

In the *working memory domain*, boys with FXS have a relative strength in verbal memory especially for recalling meaningful information with long or short delays compared to much weaker visual–spatial working memory (Munir, Cornish, & Wilding, 2000). This profile differs to that of other disorders of known genetic origin (e.g., DS and Williams syndrome) (Cornish, Wilding, & Grant, 2006; Wilding & Cornish, 2004). In the *visuo-spatial domain*, boys and girls with FXS have core deficits in their ability to navigate their environment but are proficient in their ability to process local information; a profile that differs from that of children with DS (Cornish, Munir, & Cross, 1998; Cornish, Munir, & Cross, 1999). In the *language* domain, weaknesses in *speech fluency* characterized by repetitive and impulsive speech (Belser & Sudhalter, 2001; Cornish *et al.*, 2004), are often accompanied by difficulties with grammar and pragmatics (e.g., Abbeduto *et al.*, 2004), a profile that contrasts with the language profile of Williams syndrome which is characterized by a relative proficiency in language overall, but with subtle impairments across all language subdomains (Grant, Valian, & Karmiloff-Smith, 2002; for reviews see Karmiloff-Smith & Thomas, 2003; Laing *et al.*, 2002; Thomas & Guskin, 2001).

Finally, in the *attention* domain research has produced some striking findings. Notably, the degree of inhibitory impairment displayed in FXS appears to be significantly different from that shown in children with DS, children with Williams syndrome, and typically developing comparison children (e.g., Munir *et al.*, 2000; Scerif, Cornish, Wilding, Driver, & Karmiloff-Smith, 2007; Scerif *et al.*, 2004). Using a novel computerized paradigm developed by Wilding (e.g., Wilding, Munir, & Cornish, 2001), and modified by Scerif *et al.* (2004), in which a child has to locate individual "monsters" amongst specific hidden targets (known as *visual search*), toddlers, and children with FXS show a pervasive inability to switch attention set, from a previously reinforced stimulus pattern to a new one. The degree of perseverative impairment is a *hallmark* feature of FXS and persists across the lifespan (Cornish, Scerif, *et al.*, 2007). Thus, the apparently simple operation of inhibiting the response just made in order to proceed to the next one is crucial in any chain of behavior and is disastrously impaired in FXS.

Loesch *et al.* (2003) have also reported a specific deficit in attentional control in adult FXS males; and the findings from recent studies of individuals who are carriers of FXS strongly indicate inhibitory control difficulties that cannot be accounted for by general developmental delay, because these males display IQ's within the normal range of ability (e.g., Cornish, Li, *et al.*, 2008; Grigsby *et al.*, 2008).

From a *developmental perspective*, few studies have sought to examine age trajectories in cognitive performance across different time points in the same cohort of FXS children. To date, the majority of published studies have tended to incorporate a cross-sectional design to explore age trajectories. Although such data provide important clues to possible changes in performance between cohorts of different ages, it can only ever provide a snapshot of performance in time, but, however traced, cannot provide information on developmental change. Using a longitudinal methodology, performance can be explored within a live and dynamic context that can address, for the first time, whether cognitive *profiles* in FXS change over time or whether profiles remain "frozen" after reaching a developmental plateau. Focusing on the attention and working memory domains, preliminary data from Scerif and colleagues suggests that performance although delayed with respect to chronological age and developmental level is nonetheless dynamic, not static, showing improvements over a 12-month period in the same pattern as observed in typically developing children. This profile is suggestive of developmental change not developmental freeze (e.g., Cornish, Cole, Karmiloff-Smith, & Scerif, submitted).

V. Social and Behavioral Profiles

The social and behavioral profiles of children with FXS include core difficulties in anxiety and hyperarousal and many features also overlap considerably with Autism Spectrum Disorder (ASD) and Attention Deficit Hyperactivity Disorder (ADHD) and so will be discussed alongside these two other disorders.

A. SOCIAL ANXIETY AND HYPERAROUSAL

Hypersensitivity, social anxiety, and hyperarousal are recognized as early prominent behavioral features of children with FXS and are present in both boys and girls. Even though there is a desire for social contact (Simon & Finucane, 1996; Turk & Cornish, 1998), children with FXS show social anxiety, with delay in initiating interaction, gaze avoidance, and failure to understand gaze direction (e.g., Garrett, Menon, MacKenzie, & Reiss, 2004). However, the majority of children with FXS, although tending to avoid social interactions, will offer what is now classically termed the "*FXS handshake*," whereby an initial wish to communicate socially, with a "handshake," a socially acceptable remark or even brief initial eye contact, is coupled with active and even persistent gaze avoidance. In a recent physiological study by Hall, Maynes and Reiss (2009) that involved 50 boys and girls with FXS participating in an intense "social interaction" session, the authors observed significant gaze avoidance at session onset which slightly decreased over the course of the session itself. Furthermore, eye avoidance was not associated with atypical cardiovascular activity suggesting a window of opportunity for social skill interventions. The constellation of behavioral symptoms clearly suggests that children with FXS may be overwhelmed by the demands created by social involvement, novel or unexpected situations and changes, even by the common transitions of daily life. It is not surprising therefore that huge interest has focused on the association between FXS and autism.

B. AUSTIC SPECTRUM DISORDER BEHAVIORS

There are currently very few single-gene disorders for which there is a certainty of the involvement of autism; FXS is one. Identified as one of the most common of childhood psychiatric disorders, recent studies of autism prevalence report rates of 38.9 per 10,000 in a birth cohort of children (Baird *et al.*, 2006), 18.9 per 10,000 in preschool children (Chakrabarti & Fombonne, 2005), and 20.5 per 10,000 (Gillberg, Cederlund, Lamberg, & Zeijlon, 2006). Defined by a constellation of behaviors encompassing deficits in reciprocal social interaction, communication, and repetitive, stereotyped and restricted patterns of behavior or interests, autism has intrigued researchers and clinicians for decades. Impairments in eye contact, social aloofness, and a difficulty in understanding the intentions and perspective of others can result in difficult social interactions regardless of degree of cognitive impairment.

Although still controversial, a plethora of studies using a variety of standardized measures (e.g., ADOS-G, ADI-R, CARS) indicate a range of 24–33% of FXS children will fulfill a clinical diagnosis of autism and almost all children with FXS will display some features of autism (e.g., Bailey, Hatton, Mesibov, Ament, & Skinner, 2000; Bailey, Hatton, Skinner, & Mesibov, 2001; Bailey *et al.*, 1998). However, similarities in behavioral characteristics do not always imply identical cognitive mechanisms and by default identical treatment approaches. Accumulating evidence suggests that although some "core" characteristics appear to unite FXS and autism, these same characteristics also serve very different functions and are suggestive of disparate mechanisms underlying the profiles of these two disorders. For example, *poor eye gaze* is a key characteristic of both the FXS and autism profiles and yet appears to serve quite different purposes. In autism, abnormal eye gaze is especially evident in social interactions and appears to be motivated both by a lack of understanding of the social situation itself and by the absence of a desire to communicate. In contrast, eye gaze behavior in FXS does not appear to be motivated by a lack of social awareness or a desire to communicate but is more likely due to the hyperarousal caused by social interactions. Thus despite a growing mutual relationship this eye gaze avoidance may persist. As already suggested, research suggests that FXS is associated with a unique pattern of hyperarousal and social anxiety that can cause them to avert their eyes in a social situation (to avoid the sensory stimulation of eye contact) but may still wish to communicate socially (Cornish *et al.*, 2004; Wolff, Gardner, Paccla, & Lappen, 1989).

C. ATTENTION DEFICIT HYPERACTIVITY DISORDER

Attentional problems are the most frequently cited behavioral characteristics in FXS, affecting both boys and girls but to quite different degrees of involvement depending on whether there is an associated developmental delay. However, for many children the severity of symptoms is such that it often leads to a clinical diagnosis of ADHD. In the largest nation-wide parent survey to date of children with FXS, Bailey and colleagues report findings on 976 boys and 259 girls. Of these, parents identified inattentive behaviors as a significant problem in 84% of boys and 67% of girls (Bailey, Raspa, Olmsted, & Holiday, 2008). However, further analysis revealed that of those girls who were reported as having a developmental delay, 82% had also been diagnosed with attention problems, a figure comparable with the incidence reported in males. This elevated incidence of attention problems associated with FXS is striking and further detailed assessments are needed to establish the range of inattentive behaviors across the lifespan and whether they are gender-specific or even disorder-specific. Currently, elevated rates of attention difficulties have tended only to be reported in FXS boys using both standardized rating scales (e.g., Child Behavior Checklist [CBCL]) (Cornish, Munir, & Wilding, 2001b; Hatton *et al.*, 2002; Sullivan *et al.*, 2006) and clinical interview (e.g., Parental Account of Childhood Symptoms Interview). Although at first blush the pattern of symptoms seems to mirror those of ADHD children without FXS, a classic study by Turk (1998) was the first to suggest different profiles. Turk compared the ADHD profiles of 49 FXS boys (aged 4–16 years) to that of 45 boys with DS (aged 4–16 years), and 42 boys with intellectual disability of an unknown cause (aged 4–16 years). Although both groups of boys showed similar levels of motor activity, the boys with FXS show significantly more inattentiveness, restlessness, fidgetiness, distractibility, and impulsive tendencies suggestive of DSM-IV ADHD predominantly inattentive type. These findings and others that have followed this initial study (e.g., Cornish *et al.*, 2001a) suggest that we need to question the extent to which symptom overlap implies common etiologies, or whether so-called "commonalities" in overt phenotypic behavioral outcomes actually reflect different underlying cognitive and brain processes that diverge from the normal pattern over developmental time and across syndromes. If this is the case, then intervention approaches both clinical and educational, need to be disorder-specific focusing on the biological mechanisms and their multiple manifestations at the behavioral and cognitive levels.

VI. Interventions

Alongside the tremendous growth in our new knowledge on the causes and cognitive outcomes of children with FXS there is a critical need to develop resources packages that can bridge the gap between this generation of new scientific knowledge and the uptake and utilization of these discoveries by educators, clinicians, and affected families. A possible explanation for this discrepancy may lie in the lack of available resource tools that provide up-to-date and accessible information on the range and specificity of cognitive, behavioral, and social profiles and the interventions that will promote optimal and flexible learning strategies across different environments (the home, school, community, and workplace). Resources need also to target core transition periods and be flexible enough to accommodate developmental changes such that clinicians and teachers do not to make the *a priori* assumption that what works effectively in a preschool child with FXS will be equally as effective in a teenager with FXS. It cannot be a "one size fits all" approach to intervention.

A. PSYCHOPHARMACOLOGICAL TREATMENT APPROACHES

There is no single treatment or intervention for FXS, rather treatment approaches focus on specific problems and behaviors. One approach has been to target problem behaviors that also occur in other developmental disorders for which treatment is well established. In children with ADHD, in which core symptoms overlap in FXS, the most intensive research activity has focused on the efficacy of stimulant medication to facilitate symptom reduction. There is good reason for this, notably that stimulant medications do have a markedly beneficial effect on alleviating inattentive and hyperactive behaviors by increasing dopamine levels that are thought to be reduced in ADHD. Findings from animal and human research clearly demonstrate that lower levels of dopamine in the prefrontal cortex can impact on behaviors that include hyperactivity and attentional control (e.g., Biederman & Spencer, 2008; Levy, 2008). The stimulant medication of choice, world-wide, is *methylphenidate* (MPH) which is a catecholaminergic stimulant that increases dopamine levels in the brain by blocking their reuptake. However, as with all stimulant medications there are potential adverse effects although treatment appears to be successful in reducing ADHD symptoms in the short-term (Huss, Poustka, Lehmkuhl, & Lehmkuhl, 2008) with good tolerance for its side effects, especially with careful monitoring of dosage amounts. However, evidence on the long-term effects of MPH treatment in terms of dosage and reduction in behavioral symptoms is still needed.

In FXS, despite advances in our understanding of the behavioral phenotype there is a surprising lack of empirical research on the effectiveness of psychopharmacological treatments. At the core root of these concerns are the issues surrounding safety and tolerability of medication and, in particular, the noted elevated risk of developing adverse effects to MPH above and beyond that seen in children with ADHD alone (e.g., Research Units on Pediatric Psychopharmacology Autism Network, 2005). There is also a concern that the intellectual impairment coupled with disorder-specific psychiatric or motor anomalies may produce more variable and inefficient responses to MPH than for children with ADHD symptoms in the general school population; hence, studies are relatively scarce (see Cornish & Wilding, 2010 for further exploration of these issues).

Of the few published studies on FXS, stimulant medication has been shown to improve symptoms of ADHD. For example, an early study by Hagerman and her colleagues remains one of the only studies to date to assess the effectiveness of MPH in children with FXS using a placebo, double-blind crossover trial (Hagerman, Murphy, & Wittenberger, 1988). Findings indicated that MPH was tolerated by over two-thirds of participants, with beneficial effects noted on ratings of inattentive behaviors and social skills. However, in younger children stimulants can result in symptoms of irritability and behavior problems (Hagerman et al., 2009). Given this side-effect, alternative pharmacological treatments for reducing ADHD symptoms in FXS have recently been explored. For example, Torrioli et al. (2008) used a double-blind parallel study design to assess the effect of L-acetylcarnitine (LAC; a nonstimulant agent) versus placebo in children with FXS and a dual diagnosis of ADHD aged between 6 and 13 years (mean age 9.18 years). All children were assessed at baseline and after 1, 6, and 12 months of treatment. Preliminary findings demonstrated that LAC did significantly reduce hyperactive behavior and increase attention in boys with FXS across all time points. Clearly, however, there is a critical need for more detailed investigations that assess a broad range of psychopharmacological interventions and their efficacy in treating ADHD symptoms in FXS.

In terms of other aspects the FXS phenotype, selective serotonin reuptake inhibitors (SSRIs) have been suggested as a pharmacological treatment for anxiety, and may be helpful in reducing symptoms for some individuals with FXS (50% or more of cases) (Berry-Kravis & Potanos, 2004; Hagerman et al., 2009). Recently, Hagerman and colleagues have suggested low dose antipsychotic drugs such as risperidone, may help to reduce high levels of aggression, mood instability, and severe tantrums (Hagerman et al., 2009). However, as with all pharmacological approaches to treatment, neuroleptic drugs such as risperidone can be associated with significant side effects such as dystonic reactions (e.g., tremor, muscular rigidity, and motor restlessness), and weight gain with ultimate risk of metabolic problems such as diabetes.

B. PSYCHOSOCIAL INTERVENTIONS

Alongside pharmacological interventions, behavioral and cognitive based treatment approaches can be effective in reducing symptoms of ADHD. One of the core advantages of behavioral treatments is that they address a broader range of difficulties than those directly related to the clinical symptoms of ADHD: hyperactivity, impulsivity, and distractibility. For example, the deleterious impact of ADHD on academic functioning, on social and family relationships, and on adherence to societal rules is well documented. It is therefore important that treatment, if it is to be of maximal effect, includes the child, their teachers, and parents, and that techniques can be transferable across different settings. For children with ADHD, modifying parental expectations and facilitating strategies to cope more effectively with challenges has proved a useful tool in reducing the impact of ADHD at home. In brief, such programs generally comprise weekly training sessions that aim to provide parents with skills to recognize and address problem behaviors (e.g., monitoring problem behaviors, setting rules, reinforcing positive behaviors). Antshel and Barkley (2008) provide an excellent review of this type of procedure. In recent years, an emerging body of research has begun to explore how intensive computerized training methods can improve the amount of information that children with ADHD can attend to and are able to retain and process. Klingberg *et al.* (2005); see also www.cogmed.com tested the efficacy of a 25-day computerized training program consisting of visuo-spatial tasks (referred to as "Robomemo"). Children aged 7–12 years diagnosed with ADHD were divided into treatment and control groups; the treatment group was trained with increasing levels of task difficulty while the control group remained on the lowest level throughout the training period. At the end of the training period (5–6 weeks after initial assessment), and again 3 months later, the children were tested on abilities unrelated to the training task (e.g., forward and backward spatial span, digit span). The treatment group was superior on all measures at both testing points, and parents, but not teachers, rated them as having reduced inattention and hyperactivity/impulsivity. Holmes, Gathercole and Place (2008; see also Gathercole, 2008) also assessed the efficacy of Robomemo and found substantial benefits for children with ADHD.

To what extent these techniques can be transferred to children with FXS is as yet unknown. It is therefore disappointing to see virtually no studies that have focused on the efficacy of psychosocial treatment approaches to reducing the effects of inattention and hyperactivity in atypical populations other than in children and adolescents with ADHD alone. Given that there is now clear consensus that intellectual impairment does not necessarily imply global cognitive delay, alongside the reluctance of many professionals to rely solely on medication to alleviate ADHD symptoms, it is critical that research evaluates the effectiveness of behavior modification approaches that target disorder-specific profiles both in terms of short- and long-term outcomes. In FXS, the high level of stress experienced by most parents would suggest that interventions aimed at reducing stress levels by providing behavioral techniques that lead to reductions in problem behaviors, for example, those presented by severe inattention and hyperactivity, may be extremely beneficial. Likewise, adapting computer-based interventions similar to those used in ADHD children to facilitate working memory and attentional control may be especially beneficial to children with FXS. For example, developing strategies that tap into their strength for meaningful information and their natural affinity for computers (requires no eye contact) is likely to promote considerable improvement in cognitive functioning. See Hall (2009) for a summary of current behavior treatment approaches for FXS.

C. COGNITIVE BEHAVIOR THERAPY

Despite being a primary treatment approach for anxiety problems, considerably less research has been undertaken on the efficacy of cognitive behavior therapy for anxiety in children with ID. However, some have emphasized the importance of this approach to treatment (Dagnan & Jahoda, 2006; Lindsay, 1999; Lindsay, Neilson, & Lawrenson, 1997) and although the applicability of cognitive behavior therapy as a treatment approach with children and young people with intellectual disability has been questioned, particularly for those with severe ID, there is a growing body of literature demonstrating that with modification it is possible and indeed desirable, to use these therapeutic techniques to assist people with intellectual disability (Hatton, 2002; Whitehouse, Tudway, Look, & Kroese, 2006). For example, it has been demonstrated that the ability to link thoughts and feelings (a critical skill in cognitive behavior therapy) can be taught to adults with a mild degree of ID in a single therapy session (Bruce, Collins, Langdon, Powlitch, & Reynolds, 2010), although there is some evidence that with declining verbal ability it is more difficult to successfully apply cognitive techniques (Joyce, Globe, & Moody, 2006; Sams, Collins, & Reynolds, 2006).

In contrast to the literature on cognitive behavioral approaches to treatment of mental health problems in ID, there is comparatively more evidence supporting the efficacy of this approach for the treatment of anxiety in children and adolescents with Asperger's Disorder and high functioning autism (Chalfant, Rapee, & Carroll, 2007; Lang, Regester, Lauderdale, Ashbaugh, & Haring, 2010; Moree & Davis, 2010; Sofronoff, Attwood, & Hinton, 2005; Wood *et al.*, 2009). However, there is a marked lack of controlled trials with children and young people with low functioning autism and other developmental disorders including FXS.

As already stated, it is puzzling how few early behavioral interventions are currently for children with FXS. With positive results in terms of developmental skills, language, social communication, adaptive behavior, and reduction of behavior problems in young children with autism (Corsello, 2005; Dawson *et al.*, 2010; Green *et al.*, 2010; Kasari, Freeman, & Paparella, 2006; Rogers, 1996), it is not clear why similar intervention studies have not been undertaken in FXS syndrome. Similarly, training parents of children with autism to implement treatment programs has also demonstrated gains in communicative behavior, parent communication style, reduction in behavior problems, improved parent–child interaction, and decreases in parent stress and mental health problems (McConachie & Diggle, 2005; Tonge *et al.*, 2006; Whittingham, Sofronoff, Sheffield, & Sanders, 2009). This approach to treatment delivery is well established as useful in the reduction of behavior and emotional problems in children with ID without autism (e.g., Ciechomski, Jackson, Tonge, King, & Heyne, 2001; Hudson *et al.*, 2003; Plant & Sanders, 2007). These approaches to intervention are likely to be of considerable benefit to children with FXS and their families.

In terms of enhancing social communication skills, approaches already developed for children and young people with autism are also likely to be helpful for children with FXS. The use of social stories (Lorimer, Simpson, Smith Myles, & Ganz, 2002; Moore, 2004; Reynhout & Carter, 2006; Scattone, Wilczynski, Edwards, & Rabian, 2002; Thiemann & Goldstein, 2001) and training in emotion recognition and theory of mind skills (Howlin, Baron-Cohen, & Hadwin, 1999; Ozonoff & Miller, 1995) may be helpful in improving social interactions in children and young people with FXS. Social skills training programs for children adolescents with autism have demonstrated some efficacy, namely in improving children's knowledge of social skills, increasing number of social engagements, improving greeting behaviors, improving play skills, and overall improvements in social skills (Barry *et al.*, 2003; Beaumont & Sofronoff, 2008; Krasny, Williams, Provencal, & Ozonoff, 2003; Laugeson, Frankel, Mogil, & Dillon, 2009; Laushey & Heflin, 2000). However, to date, this social skills research is limited to children with Asperger's Disorder and high functioning autism. Evaluations of these approaches to intervention in FXS are urgently needed; crucially along with the identification of characteristics which best predict successful outcomes.

In conclusion, research clearly shows that treatments to alleviate problem behaviors and improve cognitive functioning in children with ADHD and autism can be extremely effective. Although less researched there is reason to believe that similar techniques, with modifications, may be as useful in FXS. The key is for interventions to begin as early as possible in development and to use an approach that combines both a judicious use of psychopharmacological medications alongside intensive behavioral therapy.

VII. Summary and Future Directions

Infants, children, and adolescents with FXS represent a unique constellation of strengths and challenges that impact across developmental time affecting cognitive, social, and behavioral functioning. The past decade has seen tremendous advances in our understanding of this disorder and its dynamic interplay across multiple levels: the molecular, the brain, the cognitive, and the behavioral. Most recently, there has been a push in three core directions, all of which need more thorough and rich data if we are to facilitate a new generation of translational research that will be of tremendous advantage to clinicians, teachers, and affected families.

The *first* is a need for more longitudinal research that will trace developmental changes in cognitive abilities from early infancy and across different neurodevelopmental disorders. For example, our own work on FXS and DS children in the domain of *attention* has shown that despite large overall delay and greater adult difficulties with selective attention, individuals with DS showed improvements by adulthood, whereas those with FXS did not, especially for measures tapping inhibitory control (e.g., Cornish, Scerif, *et al.*, 2007). We are currently completing one of the few studies to date that have charted cognitive and behavioral inattention in FXS toddlers and young children (Scerif, Cornish & Karmiloff-Smith, funded by the Wellcome Trust, UK).

The *second* is the critical need to create an awareness that FXS does not always occur in isolation but that many children also present with symptoms that resemble more common disorders such as *ADHD* and *autism*. However, research needs to carefully investigate whether commonalities in symptomatology infer common causal mechanisms. The research to date suggests that different pathways characterize different disorders even though at first blush all share common behavior end-states, for example, ADHD symptoms. This has huge implications for treatment and for the necessity of disorder-specific interventions and approaches.

The *third* relates to treatment approaches. More specifically to the need for clinical trials (currently underway) that target problem behaviors through appropriate medications. However, there is a huge need to evaluations of behavioral and psychosocial interventions that can be used as early as possible by parents and educators in order to promote success and social inclusion across the academic trajectory. Clearly, early diagnosis of FXS is crucial if educational and clinical interventions are to have maximum impact in enabling children with FXS to develop to their maximum potential.

In conclusion, we are moving along the next decade of exciting discoveries that will ultimately demonstrate, in even finer-tuned detail, how variations in a single gene such as the *FMR1* gene result in specific cognitive and behavioral profiles and identify their time course across development.

REFERENCES

Abbeduto, L., Seltzer, M. M., Shattuck, P., Krauss, M. W., Orsmond, G., & Murphy, M. M. (2004). Psychological well-being and coping in mothers of youths with autism, Down syndrome, or fragile X syndrome. *American Journal of Mental Retardation, 109*(3), 237–254.

Ansari, D., Donlan, C., & Karmiloff-Smith, A. (2007). Typical and atypical development of visual estimation abilities. *Cortex, 43*(6), 758–768.

Antshel, K. M., & Barkley, R. (2008). Psychosocial interventions in attention deficit hyperactivity disorder. *Child & Adolescent Psychiatric Clinics of North America, 17*(2), 421–437.

Bailey, D. B., Jr., Hatton, D. D., Mesibov, G., Ament, N., & Skinner, M. (2000). Early development, temperament, and functional impairment in autism and fragile X syndrome. *Journal of Autism and Developmental Disorders, 30*(1), 49–59.

Bailey, D. B., Hatton, D. D., Skinner, M., & Mesibov, G. (2001). Autistic behavior, FMR1 protein, and developmental trajectories in young males with fragile X syndrome. *Journal of Autism and Developmental Disorders, 31*(2), 165–174.

Bailey, D. B., Jr., Mesibov, G. B., Hatton, D. D., Clark, R. D., Roberts, J. E., & Mayhew, L. (1998). Autistic behavior in young boys with fragile X syndrome. *Journal of Autism and Developmental Disorders, 28*(6), 499–508.

Bailey, D. B., Raspa, M., Olmsted, M., & Holiday, D. B. (2008). Co-occurring conditions associated with FMR1 gene variations: Findings from a national parent survey. *American Journal of Medical Genetics, 146A*(16), 2060–2069.

Baird, G., Simonoff, E., Pickles, A., Chandler, S., Loucas, T., Meldrum, D., et al. (2006). Prevalence of disorders of the autism spectrum in a population cohort of children in South Thames: The special needs and autism project (SNAP). *Lancet, 368*(9531), 210–215.

Barry, T. D., Klinger, L. G., Lee, J. M., Palardy, N., Gilmore, T., & Bodin, S. D. (2003). Examining the effectiveness of an outpatient clinic-based social skills group for high-functioning children with autism. *Journal of Autism and Developmental Disorders, 33*(6), 685–701.

Beaumont, R., & Sofronoff, K. (2008). A multi-component social skills intervention for children with Asperger syndrome: The Junior Detective Training Program. *Journal of Child Psychology and Psychiatry, 49*(7), 743–753.

Belser, R. C., & Sudhalter, V. (2001). Conversational characteristics of children with fragile X syndrome: Repetitive speech. *American Journal on Mental Retardation, 106*(1), 28–38.

Bennetto, L., & Pennington, B. F. (2002). Neuropsychology. In R. J. Hagerman & P. J. Hagerman (Eds.), *Fragile X Syndrome: Diagnosis, Treatment, and Research.* (3rd ed., pp. 206–248). Baltimore: Johns Hopkins University Press.

Berry-Kravis, E., & Potanos, K. (2004). Psychopharmacology in fragile x syndrome: Present and future. *Mental Retardation and Developmental Disabilities Research Review, 10*(1), 42–48.

Biederman, J., & Spencer, T. J. (2008). Psychopharmacological interventions. *Child and Adolescent Psychiatric Clinics of North America, 17*(2), 439–458.

Bruce, M., Collins, S., Langdon, P., Powlitch, S., & Reynolds, S. (2010). Does training improve understanding of core concepts in cognitive behaviour therapy by people with intellectual disabilites? A radomized experiment. *British Journal of Clinical Psychology, 49*, 1–13.

Casey, B. J., Soliman, F., Bath, K. G., & Glatt, C. E. (2010). Imaging genetics and development: Challenges and promises. *Human Brain Mapping, 31*(6), 838–851.

Chakrabarti, S., & Fombonne, E. (2005). Pervasive developmental disorders in preschool children: Confirmation of high prevalence. *American Journal of Psychiatry, 162*(6), 1133–1141.

Chalfant, A. M., Rapee, R., & Carroll, L. (2007). Treating anxiety disorders in children with high functioning autism spectrum disorders: A controlled trial. *Journal of Autism and Developmental Disorders, 37*, 1842–1857.

Ciechomski, L. D., Jackson, K. L., Tonge, B., King, N. J., & Heyne, D. A. (2001). Intellectual disability and anxiety in children: A group-based parent skills-training intervention. *Behaviour Change, 18*(4), 204–212.

Cornish, K. M., Bertone, A., Kogan, C., Scerif, G., & Chaudhuri, A. (2010). Linking genes to cognition: The case of fragile X syndrome. In J. A. Burack, R. M. Hodapp, G. Iarocci & E. Zigler (Eds.), *Handbook of Intellectual Disabilties and Development.* (2nd ed.). New York, NY: Oxford University Press.

Cornish, K., Levitas, A., & Sudhalter, V. (2007). Fragile X syndrome: The journey from genes to behavior. In M. M. Mazzocco & J. L. Ross (Eds.), *Neurogenetic Developmental Disorders: Variation of Manifestation in Childhood.* Massachsetts: MIT Press. p. 73.

Cornish, K. M., Li, L., Kogan, C., Jacquemont, S., Turk, J., Dalton, A., *et al.* (2008). Age-dependent cognitive changes in carriers of the fragile X Syndrome. *Cortex, 44*(6), 628–636.

Cornish, K. M., Munir, F., & Cross, G. (1998). The nature of the spatial deficit in young females with fragile-X syndrome: A neuropsychological and molecular perspective. *Neuropsychologia, 36*(11), 1239–1246.

Cornish, K. M., Munir, F., & Cross, G. (1999). Spatial cognition in males with fragile-X syndrome: Evidence for a neuropsychological phenotype. *Cortex, 35*(2), 263–271.

Cornish, K. M., Munir, F., & Wilding, J. (2001a). A neuropsychological and behavioural profile of attention deficits in fragile X syndrome. *Revista de Neurologia, 33*, S24.

Cornish, K. M., Munir, F., & Wilding, J. (2001b). Specifying the attention deficit in Fragile X syndrome. *Revista Neurology, 33*, s24–29.

Cornish, K., Scerif, G., & Karmiloff-Smith, A. (2007). Tracing syndrome-specific trajectories of attention across the lifespan. *Cortex, 43*(6), 672–685.

Cornish, K., Sudhalter, V., & Turk, J. (2004). Attention and language in fragile X. *Mental Retardation and Developmental Disabilities Research Reviews, 10*(1), 11–16.

Cornish, K., Turk, J., & Hagerman, R. (2008). The fragile X continuum: New advances and perspectives. *Journal of Intellectual Disability Research, 52*(6), 469–482.

Cornish, K. M., & Wilding, J. (2010). *Attention, genes and developmental disorders.* Oxford: Oxford University Press.

Cornish, K. M., Wilding, J., & Grant, C. (2006). Deconstructing working memory in developmental disorders of attention. In S. Pickering (Ed.), *Working Memory and Education.* Burlington, MA: Elsevier Science.

Cornish, K. M., Cole, V., Karmiloff-Smith, A., & Scerif, G. (submitted). Mapping trajectories of attention and working memory in young children with fragile X syndrome: Developmental freeze or developmental change?.

Corsello, C. (2005). Early intervention in autism. *Infants and Young Children, 18*(2), 74–85.

Dagnan, D., & Jahoda, A. (2006). Cognitive–behavioural intervention for people with intellectual disability and anxiety disorders. *Journal of Applied Research in Intellectual Disabilities, 19,* 91–97.

Dawson, G., Rogers, S., Munson, J., Smith, M., Winter, J., Greenson, J., *et al.* (2010). Randomized, controlled trial of an intervention for toddlers with autism: The Early Start Denver Model. *Pediatrics, 125*(1), e17–e23.

Eliez, S., Blasey, C. M., Freund, L. S., Hastie, T., & Reiss, A. L. (2001). Brain anatomy, gender and IQ in children and adolescents with fragile X syndrome. *Brain, 124*(8), 1610–1618.

Garrett, A. S., Menon, V., MacKenzie, K., & Reiss, A. L. (2004). Here's looking at you, kid: Neural systems underlying face and gaze processing in fragile X Syndrome. *Archives of General Psychiatry, 61*(3), 281–288.

Gathercole, S. (2008). Working memory in the classroom. *The Psychologist, 21*(5), 382–385.

Gillberg, C., Cederlund, M., Lamberg, K., & Zeijlon, L. (2006). Brief report: "The autism epidemic". The registered prevalence of autism in a Swedish urban area. *Journal of Autism and Developmental Disorders, 36,* 429–435.

Grant, J., Valian, V., & Karmiloff-Smith, A. (2002). A study of relative clauses in Williams syndrome. *Journal of Child Language, 29*(2), 403–416.

Green, J., Charman, T., McConachie, H., Aldred, C., Slonims, V., Howlin, P., *et al.* (2010). Parent-mediated communication-focused treatment in children with autism (PACT): A randomised controlled trial. *Lancet, 375*(9732), 2152–2160.

Grigsby, J., Brega, A. G., Engle, K., Leehey, M. A., Hagerman, R. J., Tassone, F., *et al.* (2008). Cognitive profile of fragile X premutation carriers with and without fragile X-associated tremor/ataxia syndrome. *Neuropsychology, 22*(1), 48–60.

Hagerman, P. J. (2008). The fragile X prevalence paradox. *Journal of Medical Genetics, 45,* 498–499.

Hagerman, R. J., Berry-Kravis, E., Kaufman, W. E., Ono, M. Y., Tartaglia, N., Lachiewicz, A., *et al.* (2009). Advances in the treatment of fragile X Syndrome. *Pediatrics, 123,* 378–390.

Hagerman, R. J., Murphy, M. A., & Wittenberger, M. D. (1988). A controlled trial of stimulant medication in children with the fragile X syndrome. *American Journal of Medical Genetics, 30*(1–2), 377–392.

Hall, S. (2009). Treatments for fragile X syndrome: A closer look at data. *Developmental Disabilities Research Review, 15*(4), 353–360.

Hall, S. S., Maynes, N. P., & Reiss, A. L. (2009). Using percentile schedules to increase eye contact in children with fragile X syndrome. *Journal of Applied Behavioral Analysis, 42*(1), 171–176.

Hatton, C. (2002). Psychosocial interventions for adults with intellectual disabilites and mental health problems: A review. *Journal of Mental Health, 11*(4), 357–373.

Hatton, D. D., Hooper, S. R., Bailey, D. B., Skinner, M. L., Sullivan, K. M., & Wheeler, A. (2002). Problem behavior in boys with fragile X syndrome. *American Journal of Medical Genetics, 108*(2), 105–116.

Hoeft, F., Hernandez, A., Parthasarathy, S., Watson, C. L., Hall, S. S., & Reiss, A. L. (2007). Fronto-striatal dysfunction and potential compensatory mechanisms in male adolescents with fragile X syndrome. *Human Brain Mapping*, *28*(6), 543–554.

Holmes, J., Gathercole, S. E., & Place, M. (2008). Working memory deficits can be overcome: Impacts of training and medication on working memory in children with ADHD, submitted for publication.

Howlin, P., Baron-Cohen, S., & Hadwin, J. (1999). *Teaching children with autism to mindread*. Chichester: John Wiley & Sons.

Hudson, A. M., Matthews, J., Gavidia-Payne, S., Cameron, C., Mildon, R., Radler, G., *et al.* (2003). Evaluation of an intervention system for parents of children with intellectual disability and challenging behaviour. *Journal of Intellectual Disability Research*, *47*(4–5), 238–249.

Huss, M., Poustka, F., Lehmkuhl, G., & Lehmkuhl, U. (2008). No increase in long-term risk for nicotine use disorders after treatment with methylphenidate in children with attention-deficit/hyperactivity disorder (ADHD): Evidence from a non-randomised retrospective study. *Journal of Neural Transmission*, *115*(2), 335–339.

Joyce, T., Globe, A., & Moody, C. (2006). Assessment of the component skills for cognitive therapy in adults with intellectual disability. *Journal of Applied Research in Intellectual Disabilities*, *19*, 17–23.

Karmiloff-Smith, A. (2009). Nativism versus neuroconstructivism: Rethinking the study of developmental disorders. *Developmental Psychology*, *45*(1), 56–63.

Karmiloff-Smith, A., & Thomas, M. (2003). What can developmental disorders tell us about the neurocomputational constraints that shape development? The case of Williams syndrome. *Development and Psychopathology*, *15*(4), 969–990.

Karmiloff-Smith, A., Thomas, M., Annaz, D., Humphreys, K., Ewing, S., Brace, N., *et al.* (2004). Exploring the Williams syndrome face-processing debate: The importance of building developmental trajectories. *Journal of Child Psychology and Psychiatry*, *45*(7), 1258–1274.

Karmiloff-Smith, A., Tyler, L. K., Voice, K., Sims, K., Udwin, O., Howlin, P., *et al.* (1998). Linguistic dissociations in Williams syndrome: Evaluating receptive syntax in on-line and off-line tasks. *Neuropsychologia*, *36*(4), 343–351.

Kasari, C., Freeman, S., & Paparella, T. (2006). Joint attention and symbolic play in children with autism: A randomized controlled intervention study. *Journal of Child Psychology and Psychiatry*, *47*(6), 611–620.

Kates, W. R., Abrams, M. T., Kaufmann, W. E., Breiter, S. N., & Reiss, A. L. (1997). Reliability and validity of MRI measurement of the amygdala and hippocampus in children with fragile X syndrome. *Psychiatry Research: Neuroimaging*, *75*(1), 31–48.

Klingberg, T., Fernell, E., Olesen, P. J., Johnson, M., Gustafsson, P., Dahlstrom, K., *et al.* (2005). Computerized training of working memory in children with ADHD - A randomized, controlled trial. *Journal of the American Academy of Child and Adolescent Psychiatry*, *44*(2), 177–186.

Krasny, L., Williams, B. J., Provencal, S., & Ozonoff, S. (2003). Social skills interventions for the autism spectrum: Essential ingredients and a model curriculum. *Child & Adolescent Psychiatric Clinics of North America*, *12*(1), 107–122.

Kwon, H., Menon, V., Eliez, S., Warsofsky, I. S., White, C. D., Dyer-Friedman, J., *et al.* (2001). Functional neuroanatomy of visuospatial working memory in fragile X syndrome: Relation to behavioral and molecular measures. *American Journal of Psychiatry*, *158*(7), 1040–1051.

Lachiewicz, A. M., Dawson, D. V., & Spiridigliozzi, G. A. (2000). Physical characteristics of young boys with fragile X syndrome: Reasons for difficulties in making a diagnosis in young males. *American Journal of Medical Genetics, 92*(4), 229–236.

Laing, E., Butterworth, G., Ansari, D., Gsodl, M., Longhi, E., Panagiotaki, G., *et al.* (2002). Atypical development of language and social communication in toddlers with Williams syndrome. *Developmental Science, 5*(2), 233–246.

Lang, R., Regester, A., Lauderdale, S., Ashbaugh, K., & Haring, A. (2010). Treatment of anxiety in autism spectrum disorders using cognitive behaviour therapy: A systematic review. *Developmental Rehabilitation, 13*(1), 53–63.

Laugeson, E. A., Frankel, F., Mogil, C., & Dillon, A. R. (2009). Parent-assisted social skills training to improve friendships in teens with autism spectrum disorders. *Journal of Autism and Developmental Disorders, 39*, 596–606.

Laushey, K. M., & Heflin, L. J. (2000). Enhancing social skills of kindergarten children with autism through the training of multiple peers as tutors. *Journal of Autism and Developmental Disorders, 30*(3), 183–193.

Levy, F. (2008). Pharmacological and therapeutic directions in ADHD: Specificity in the PFC. *Behavioral and Brain Functions, 4*, 12.

Lindsay, W. R. (1999). Cognitive therapy. *Psychologist, 12*(5), 238–241.

Lindsay, W., Neilson, C., & Lawrenson, H. (1997). Cognitive–behaviour therapy for anxiety in people with learning disabilities. In B. S. Kroese, D. Dadnan & K. Loumidis (Eds.), *Cognitive–Behaviour Therapy for People with Learning Disabilities*. London: Routledge. pp. 124–140.

Loesch, D. Z., Bui, M. Q., Grigsby, J., Butler, E., Epstein, J., Huggins, R. M., *et al.* (2003). Effect of the fragile X status categories and the FMRP levels on executive functioning in fragile X males and females. *Neuropsychology, 17*(4), 646–657.

Lorimer, P. A., Simpson, R. L., Smith Myles, B., & Ganz, J. B. (2002). The use of social stories as a preventative behavioral intervention in a home setting with a child with autism. *Journal of Positive Behavior Interventions, 4*(1), 53–60.

McConachie, H., & Diggle, T. (2005). Parent implemented early intervention for young children with autism spectrum disorder: A systematic review. *Journal of Evaluation in Clinical Practice, 13*, 120–129.

Menon, V., Leroux, J., White, C. D., & Reiss, A. L. (2004). Frontostriatal deficits in fragile X syndrome: Relation to FMR1 gene expression. *Proceedings of the National Academy of Sciences of the United States of America, 101*(10), 3615–3620.

Moore, P. S. (2004). The use of social stories in a psychology service for children with learning disabilities: A case study of a sleep problem. *British Journal of Learning Disabilities, 32*(3), 133–138.

Moree, B. N., & Davis, T. E. (2010). Cognitive–behavioral therapy for anxiety in children diagnosed with autism spectrum disorders: Modification trends. *Research in Autism Spectrum Disorders, 4*, 346–354.

Mostofsky, S. H., Mazzocco, M. M., Aakalu, G., Warsofsky, I. S., Denckla, M. B., & Reiss, A. L. (1998). Decreased cerebellar posterior vermis size in fragile X syndrome: Correlation with neurocognitive performance. *Neurology, 50*(1), 121–130.

Munir, F., Cornish, K. M., & Wilding, J. (2000). Nature of the working memory deficit in fragile-X syndrome. *Brain and Cognition, 44*(3), 387–401.

Ozonoff, S., & Miller, J. N. (1995). Teaching theory of mind: A new approach to social skills training for individuals with autism. *Journal of Autism and Developmental Disorders, 25*(4), 415–433.

Plant, K. M., & Sanders, M. R. (2007). Reducing problem behavior during care-giving in families of preschool-aged children with developmental disabilities. *Research in Developmental Disabilities*, *28*(4), 362–385.

Purser, H. R., & Jarrold, C. (2005). Impaired verbal short-term memory in Down syndrome reflects a capacity limitation rather than atypically rapid forgetting. *Journal of Experimental Child Psychology*, *9*(1), 1–23.

Reiss, A. L., Aylward, E., Freund, L. S., Joshi, P. K., & Bryan, R. N. (1991). Neuroanatomy of fragile X syndrome: The posterior fossa. *Annals of Neurology*, *29*(1), 26–32.

Reiss, A. L., Lee, J., & Freund, L. (1994). Neuroanatomy of fragile X syndrome: The temporal lobe. *Neurology*, *44*(7), 1317–1324.

Research Units on Pediatric Psychopharmacology Autism Network, (2005). Randomized, controlled, crossover trial of methylphenidate in pervasive developmental disorders with hyperactivity. *Archives of General Psychiatry*, *62*(11), 1266–1274.

Reynhout, G., & Carter, M. (2006). Social stories for children with disabilities. *Journal of Autism and Developmental Disorders*, *36*(4), 445–469.

Rivera, S. M., Menon, V., White, C. D., Glaser, B., & Reiss, A. L. (2002). Functional brain activation during arithmetic processing in females with fragile X Syndrome is related to FMR1 protein expression. *Human Brain Mapping*, *16*(4), 206–218.

Rogers, S. J. (1996). Early intervention in autism. *Journal of Autism and Developmental Disorders*, *26*(2), 243–246.

Sams, K., Collins, S., & Reynolds, S. (2006). Cognitive therapy abilities in people with learning disabilities. *Journal of Applied Research in Intellectual Disabilities*, *19*, 25–33.

Scattone, D., Wilczynski, S. M., Edwards, R. P., & Rabian, B. (2002a). Decreasing disruptive behaviors of children with autism using social stories. *Journal of Autism and Developmental Disorders*, *32*(6), 535–543.

Scerif, G., Cornish, K., Wilding, J., Driver, J., & Karmiloff-Smith, A. (2004). Visual search in typically developing toddlers and toddlers with fragile X or Williams syndrome. *Developmental Science*, *7*(1), 116–130.

Scerif, G., Cornish, K., Wilding, J., Driver, J., & Karmiloff-Smith, A. (2007). Delineation of early attentional control difficulties in fragile X syndrome: Focus on neurocomputational changes. *Neuropsychologia*, *45*(8), 1889–1898.

Simon, E. W., & Finucane, B. M. (1996). Facial emotion identification in males with fragile X syndrome. *American Journal of Medical Genetics*, *67*(1), 77–80.

Skinner, M., Hooper, S., Hatton, D. D., Roberts, J., Mirrett, P., Schaaf, J., *et al.* (2005). Mapping nonverbal IQ in young boys with fragile X syndrome. *American Journal of Medical Genetics A*, *132*(1), 25–32.

Sofronoff, K., Attwood, T., & Hinton, S. (2005). A randomized controlled trial of a CBT intervention for anxiety in children with Asperger syndrome. *Journal of Child Psychology and Psychiatry*, *46*, 1152–1160.

Song, H. R., Lee, K. S., Li, Q. W., Koo, S. K., & Jung, S. C. (2003). Identification of cartilage oligomeric matrix protein (COMP) gene mutations in patients with pseudoachondroplasia and multiple epiphyseal dysplasia. *Journal of Human Genetics*, *48*(5), 222–225.

Sullivan, K., Hatton, D., Hammer, J., Sideris, J., Hooper, S., Ornstein, P., *et al.* (2006). ADHD symptoms in children with FXS. *American Journal of Medical Genetics Part A*, *140A*(21), 2275–2288.

Tamm, L., Menon, V., Johnston, C. K., Hessl, D. R., & Reiss, A. L. (2002). fMRI study of cognitive interference processing in females with fragile X syndrome. *Journal of Cognitive Neuroscience*, *14*(2), 160–171.

Thiemann, K. S., & Goldstein, H. (2001). Social stories, written text cues, and video feedback: Effects on social communication of children with autism. *Journal of Applied Behavior Analysis, 34*(4), 425–446.

Thomas, M. S., Annaz, D., Ansari, D., Scerif, G., Jarrold, C., & Karmiloff-Smith, A. (2009). Using developmental trajectories to understand developmental disorders. *Journal of Speech Language & Hearing Research, 52*(2), 336–358.

Thomas, J. M., & Guskin, K. A. (2001). Disruptive behavior in young children: What does it mean? *Journal of the American Academy of Child & Adolescent Psychiatry, 40*(1), 44–51.

Tonge, B., Brereton, A. V., Kiomall, M., Mackinnon, A., King, N. M., & Rinehart, N. (2006). Effects on parental mental health of an education and skills training program for parents of young children with autism: A randomized controlled trial. *Journal of the American Academy of Child and Adolescent Psychiatry, 45*(5), 561–569.

Torrioli, M. G., Vernacotola, S., Peruzzi, L., Tabolacci, E., Mila, M., Militerni, R., *et al.* (2008). A double-blind, parallel, multicenter comparison of L-acetylcarnitine with placebo on the attention deficit hyperactivity disorder in fragile X syndrome boys. *American Journal of Medical Genetics, 146*(7), 803–812.

Turk, J. (1998). Fragile X syndrome and attentional deficits. *Journal of Applied Research in Intellectual Disabilities, 11*(3), 175–191.

Turk, J., & Cornish, K. (1998). Face recognition and emotion perception in boys with fragile-X syndrome. *Journal of Intellectual Disability Research, 42*(6), 490–499.

Whitehouse, R. M., Tudway, J. A., Look, R., & Kroese, B. S. (2006). Adapting individual psychotherapy for adults with intellectual disabilities: A comparative review of the cognitive–behavioural and psychodynamic literature. *Journal of Applied Research in Intellectual Disabilities, 19*, 55–65.

Whittingham, K., Sofronoff, K., Sheffield, J., & Sanders, M. R. (2009). Stepping stones triple P: An RCT of a parenting program with parents of a child diagnosed with an autism spectrum disorder. *Journal of Abnormal Child Psychology, 37*(37), 469–480.

Wilding, J., & Cornish, K. (2004). Attention and executive functions in children with Down syndrome. In J. Malard (Ed.), *Focus on Down Syndrome Research.* New York, NY: Nova Publishing.

Wilding, J., Munir, F., & Cornish, K. (2001). The nature of attentional differences between groups of children differentiated by teacher ratings of attention and hyperactivity. *British Journal of Psychology, 92*(2), 357–371.

Wolff, P. H., Gardner, J., Paccla, J., & Lappen, J. (1989). The greeting behavior of fragile X males. *American Journal of Mental Retardation, 93*(4), 406–411.

Wood, J. J., Drahota, A., Sze, K., Har, K., Chiu, A., & Langer, D. A. (2009). Cognitive behavioral therapy for anxiety in children with autism spectrum disorders: A randomized, controlled trial. *Journal of Child Psychology and Psychiatry, 50*(3), 224–234.

Author Index

A

Aakalu, G., 214
Aaron, P.G., 84
Abbeduto, L., 134–135, 138, 218
Abrams, M.T., 214
Ackerman, P.L., 58, 60, 63
Adams, A.-M., 5, 7, 11, 16–17, 59
Adams, C., 120, 145
Adams, J.W., 4, 102
Adlof, S.M., 83, 89–90, 98, 101
Adolphs, R., 183
Ahonen, T., 12
Albano, A.M., 196
Alberman, E., 132
Aldred, C., 227
Alexander, P., 117
Alku, P., 133
Allanson, J.E., 164
Allington, R.L., 110, 117
Alloway, T.P., 1–3, 5–11, 13–18, 21, 27–29, 31–32, 59, 193
Almazan, M., 168–169, 172
Altman, D.G., 116
Altmann, G., 89–90, 97
Alton, S., 153
Ament, N., 221
Amiel-Tison, C., 167
Anderson, J.R., 2
Anderson, R.C., 115, 135
Annaz, D., 175–176, 217
Ansari, D., 56, 168, 179, 184, 192, 217, 219
Antell, S.E., 49
Antonarakis, S.E., 132
Antshel, K.M., 225
Appelbaum, L.G., 176, 183
Appleton, M., 154
Archer, T., 142
Archibald, L.M., 11–12
Arnell, K.M., 64
Arnsten, A.F., 31

Arnup, J.S., 53
Aronen, E.T., 16, 18
Arsten, A.F.T., 31
Ashbaugh, K., 227
Ashcraft, M.H., 12, 55
Atkinson, D., 164
Atkinson, J., 180
Attwood, T., 227
Ayali, M., 47–48
Aylward, E., 214

B

Back, E., 186
Baddeley, A.D., 2–3, 11–12, 60, 86, 89, 102, 136, 138, 167, 170, 173
Badichi, N., 47–48
Bailey, D.B., 221–222
Bailey, D.H., 46, 51, 57–58, 61–62, 64, 68–69
Baird, G., 170, 221
Baker-Ward, L., 21
Bandettini, P.A., 31
Barbaresi, W.J., 47–48, 67
Barkley, R., 225
Barnes, M.A., 55–56, 89–90, 100, 103
Barnett, R., 12
Barnett, S., 66–67
Baron-Cohen, S., 120–121, 184, 186, 227
Barrat-Boyes, B., 163
Barrett, M.D., 135
Barrouillet, P., 2, 56–57
Barry, T.D., 227
Bartfai, A., 26
Bates, E., 168
Bath, K.G., 214
Baumann, J.F., 115
Baumeister, R.F., 17
Bawden, H.N., 196
Baylis, P., 139, 144, 151
Beadman, J., 135, 155
Beaumont, R., 227

H

Hayiou-Thomas, M.E., 48
Haynes, J., 117
Healy, A.F., 2
Heflin, L.J., 227
Hegarty, M., 2, 7, 14
Helenius, P., 108
Hellawell, D., 175
Helmkay, A., 133
Henry, L.A., 102
Hepburn, S., 170
Herman, E., 121
Herman, P.A., 135
Hernandez, A., 215
Hesketh, L.J., 11–12, 133–134
Hesselink, J.R., 133
Hessl, D.R., 215
Hewes, A.K., 12, 167, 173
Heyne, D.A., 227
Hickok, G., 164–165, 167, 183
Hilden, K., 110
Hill, G., 145
Hill, J., 186
Hill, K., 183
Hilton, K.A., 3, 8–11, 13–18, 27–29, 31–32, 66
Hinton, S., 227
Hirvikoski, T., 26
Hitch, G.J., 2–3, 12, 47, 60–61, 136
Hoard, M.K., 4, 12, 46–47, 50–51, 53–62, 64, 67–69
Hodapp, R.M., 182, 190
Hoeft, F., 108, 182, 215
Hoff, E., 7
Hoffman, J.E., 172–173
Hoffman, J.V., 83
Hogan, T.P., 81
Hogg-Johnson, S., 12–13, 16, 27–28
Holiday, D.B., 222
Holliday, R.E., 89–90
Holmes, J., 1, 3–4, 8–11, 13–18, 21, 27–32, 66, 69, 225
Hooper, S.R., 218, 222
Hoover, W.A., 81
Horowitz, T.S., 121
Howard, S., 170
Howerter, A., 9
Howlin, P., 165–166, 168–169, 177–182, 217, 227
Hudson, A.M., 227
Huggins, R.M., 219

Hughes, C., 121, 137
Hulme, C., 23, 82–83, 90, 98, 109, 116, 118, 134, 136, 140–141, 155, 178
Hulslander, J.L., 13, 16
Humphreys, K., 175–176, 217
Humphries, T., 13, 16
Hunt, A., 18, 22–23
Hunt, E., 62
Husebye-Hartman, E., 105
Huss, M., 223
Huston, A.C., 46, 64–65, 68
Hutchison, S., 24
Huttenlocher, J., 7, 49, 54
Hutton, T.J., 164

I

Iacono, T., 140–141
Idol, L., 117
Igács, J., 179
Ignace, H., 121

J

Jaccard, J.J., 84
Jackson, K.L., 227
Jacobaeus, H., 26
Jacobsen, K.C., 34
Jacobsen, S.J., 47–48, 67
Jacquemont, S., 219
Jaeggi, S.M., 24
Jahoda, A., 226
Jain, U., 28
Janosky, J., 7, 34
Janssen, R., 61
Jarrold, C., 2, 12, 136, 138, 140, 167–168, 170, 172–173, 217
Järvinen-Pasley, A., 183
Jarvis, H.L., 4
Jawaid, A., 181, 194
Jaywitz, P.B., 112
Jeffcock, E., 6, 9–10, 16–18
Jeffries, N.O., 63
Jeffries, S.A., 11
Jenkins, E., 54
Jensen, A.R., 59, 62
Jerman, O., 11, 61
Jernigan, T.L., 133, 165, 168–169, 183
Jiang, Z.D., 133

X

Xian, H., 34
Xu, F., 179

Y

Yama, A., 183
Yao, Y., 53–54
Yin, R.K., 175
York, T.P., 145
Young, A.W., 175
Ypsilanti, A., 134

Yuill, N., 100, 102, 104, 111–112, 117
Yule, W., 166

Z

Zacks, R.T., 91
Zeijlon, L., 221
Zhang, X.Y., 7
Zheng, X., 61
Zhu, X.L., 164
Zijdenbos, A., 63

Subject Index

A

Academic skills, WS
  literacy
    mental age equivalence scores, 177
    reading abilities, 178
    standardized scores, 177–178
    verbal and nonverbal intellect, 177
  numeracy, 179–180
  spelling, 178–179
ADHD, *see* Attention deficit hyperactivity
    disorder
Arithmetic, MLD
  development
    counting, 54
    direct retrieval and decomposition,
      54–55
  and low achieving (LA) children
    fact retrieval, 56–58
    procedural competence, 55–56
ASD, *see* Autism spectrum disorder
Attention deficit hyperactivity disorder
    (ADHD); *see also* Behavioral profile,
    poor comprehender
  behavior
    profiles, 17
    symptoms treatment, 28
  behavioral and cognitive based treatment,
    225–226
  characterization, 12–13
  childhood, characterization, 16–17
  domain-general, 14
  FXS
    behavioral and cognitive based
      treatment, 225–226
    symptoms and diagnosis, 222–224
  inattentive and distractible behavior,
    13–14
  reading comprehension skills, 92

training effects, 26
visuospatial STM, 12–13
working memory
  deficits and inattentiveness, 16–17
  profiles, 13
WS, 180, 196
Autism spectrum disorder (ASD), 120–121

B

BAS II, *see* British ability scales II
Behavioral profile, poor comprehender
  assessment
    individual's own experience, 87–88
    material, reading, 87
    multiple choice, 89
    questions and response methods, 88–89
    reading comprehension, 86–87
    tests, 87
  criteria, reading
    age-appropriate levels and accuracy, 91
    comprehension and accuracy, 90
    cutoff, comprehension, 89
    vocabulary, 90–91
  difficulties
    ADHD children, 92
    classroom, 91
Benton facial recognition test, 175
Biological profile, poor comprehender
  genetic
    family, 106–107
    twin sample, 107
  neural
    comprehension and phonological
      deficits, 108
    functional neuroanatomy, 107–108
    neuroimaging studies, functional, 108
    underpinnings, 108–109
Boston naming tests, 168

Contents of Previous Volumes